Nanoart

# Nanoart
## The Immateriality of Art

Paul Thomas

**intellect** Bristol, UK / Chicago, USA

First published in the UK in 2013 by
Intellect, The Mill, Parnall Road, Fishponds, Bristol, BS16 3JG, UK

First published in the USA in 2013 by
Intellect, The University of Chicago Press, 1427 E. 60th Street,
Chicago, IL 60637, USA



A catalogue record for this book is available from the British Library.

Cover designer: Ellen Thomas
Copy-editor: MPS Technologies
Production manager: Bethan Ball
Typesetting: Planman Technologies

ISBN: 978-1-84150-708-8

# Contents

## Foreword

Although nanotechnology has been with us for several decades in hard science, it still has the aura of an alluring speculative fiction with plenty of room for imaginative movement. History has it that Richard Feynman famously postulated the prospect of nanotechnology, although without naming it as such, during a lighthearted after-dinner speech to scientific colleagues at Caltech way back in 1959. His entertaining thought-experiment, dealing with the manipulation of individual atoms in manufacturing and with the fabrication at an atomic level of otherwise unimaginably miniaturized and complex electrical circuitry for computing, was quirkily titled 'There's Plenty of Room at the Bottom'. 'A new field of physics', declared the journal which, within little more than a month, had published this potentially epochal – certainly visionary – talk, and a 'fascinating field', they added, in which 'Dr Feynman … issues an invitation to all scientists – and non-scientists – to start working'.[1] A generous invitation, although the publisher's parenthetical inclusion of 'non-scientists' does not suggest an ecumenical reconciliation of what C.P. Snow had momentously declared, at Cambridge earlier in the year of Feynman's talk, as divorce between 'the two cultures'. The non-scientists of the journal's editors would probably not have been otherwise excluded envoys from the humanities, but the very much included corporate sponsors – military and industrial manufacturers – whose full pages of grandiloquent advertorial liberally interrupt Feynman's text.[2]

Given the informal and somewhat conjectural incitements of this text, the role of Feynman's talk in inaugurating the history of nanotech may be more legendary than actually catalytic.[3] But precisely due to its slightly fanciful edge, Feynman's talk can put a slightly different spin on the invitation to the 'non-scientist', different from that of his editors' solicitation. And it happens with an unremarked minor detail in his narrative description of the technology; a little detail that has the intrigue of a symptomatic slip-up. Not an error (one would not dare charge Feynman with that), but a momentary slide or swerve on a rhetorical banana-skin that turns us in quite another direction to that of the military-industrial complex which appeared to provide such solid foundations for the exploitable potential of nanotech. It turns us towards art.

Picturing the process for this atomic scaled engineering, Feynman nonchalantly (and, he adds, 'for amusement' [Feynman 1960: 26]) introduced an analogy with a mechanical drafting tool – one which, disarmingly, originates in the seventeenth century – called the pantograph: a device constructed from rods hinged in a parallelogram, rather like the common 'concertina' extension arm on adjustable wall-mounted bathroom mirrors (Feynman 1960: 30, 34). In its conventional and simplest mode, the operator would trace the outlines of a form with a stylus positioned at one of the pantograph's hinged junctions. The pantograph would then automatically copy that trace through the parallel motion of a marker pencil located on another of the apparatus's arms. Depending on the location of that marker pencil, this copy could be an equivalent size replica or – demonstrating the real ingenuity of the mechanism – it could produce an exact but either magnified or reduced facsimile.

Feynman evidently had in mind 3D milling machines run by pantographs on servo-motors rather than manual drafting table ones, and so he referred to a device with more industrial than artistic guise. But it is of no less interest – in fact, it is more tantalizing as a conceit worthy of a steampunk novel – that this celebrated advent of nanotech involved a rather clunky visual simile that conjures up the artist (or at least a copyist in a monastic *scriptorium)* as well as the engineer. Indeed as a verbal image, Feynman's illustration could be reminiscent of the artist El Lissitzky's justly famous photomontage from 1924, 'The Constructor', in which a navigator's or geometrician's pair of compasses is deftly poised in the artist's open hand, which is superimposed over his eye. El Lissitzky's iconic self-portrait as a kind of artistic pilot (naval or aeronautical, charting the conjoined directions of his society and art) is a cocktail of persuasive boast and artistic manifesto. Incorporating his own innovative stencil typography to replace Russian Cyrillic, Lissitzky's portrait is also an advertisement for a visionary social project of expanded literacy as well a manifesto for his revolutionary applications of the photographic medium.

The apparatus in Feynman's talk likewise conjures up an operator working with revolutionary promise at unprecedented magnitudes: not simply with cute Lilliputian minutiae but with a vastness of possibility where ratios of labour to material, material stresses to performance, and performance to equipment bulk, blow out.[4] At the atomic level, there is, for instance, much more empty volume than matter, and so there is 'plenty of room'. Anachronistic as it may sound, there is nothing naïve about the science implied by Feynman's pantograph analogy. Because the scope of such a pantograph would have to extend beyond mere miniaturization, Feynman acknowledges that the translation from a microscopic dimension to the nano-scale would entail a phase transition and require remediation through tools which were themselves nanometric; not just a continuously graduated scale of reduction from the hand or eye.

Feynman's pantograph acknowledged a gulf between the agency articulating the mechanism and the sort of tools needed to move atoms around. It also acknowledged that any industry conducted at the atomic level (specifically, that would mean below 100 nanometres) would involve proportions of natural forces different to those at scales larger than molecular configurations. Gravity, weight and inertial drag for instance, would play diminished roles; but the complex of (comparatively, at macro scales) weak van der Waals forces – repulsive and attractive intermolecular and intramolecular energies – would become dominant, and consequently would expose the technological diagrams and processes of atomic level manufacture to the bizarre topography and capricious vivacity of the quantum realm.

If we were to take Feynman at his word – and we would be both impertinent and unwise *not* to do so – how could we envisage such a simple mechanical instrument as his pantograph extended in its range enough to achieve a veritable 'quantum leap' into the nanosphere? If we do take this visual analogue literally, is it as foolish as imagining the number of angels that can dance on the head of a pin?[5] Actually, both this literalism of the imagination and this folly can precisely be claimed as the province of artists. It is a charming coincidence that at the start of Paul Thomas's book charting his professional artistic reflections on the nanosphere – deliberations inflected by creative and pragmatic acquaintance with that realm's weird physics – we find a puzzle dramatically pertinent to the pantograph's activity of drafting. Thomas begins his book by speculating on the nanometric interactions between the graphite in a pencil and the surface of the paper on which that pencil draws.

It is not the macroscopic dimension of the mark that holds his attention here (which would be the domain of artistic discipline, the connoisseurship of style, art history, studio pedagogy, perhaps of media theory), but the tangled and fluctuating realm below the visibility of that mark and indeed beneath its disciplined significance, the fluxes of which lead him along various paths from Lucretius's atomism to Bergsonian vitalism to the motion blur and simultaneous temporal facets of Cubo-Futurism; from a phenomenological continuum to the ever-receding discontinuous incidents of Zeno's arrow. Diverse and incommensurate as some of these notions are, it is the 'mark' – the touch or facture – that initializes all these movements of enquiry in Thomas's book.

'Mark-making', by the way, is jargon that has had an odd resurgence among many contemporary visual artists these days. Its older, twentieth-century usage alluded to the urgent signature-effect and totemic symbolization of painterly gestures in the legacy of Abstract Expressionism. More recently, 'marking' has had an urban and subcultural charge identified with the graffitist's tag and in the inscriptions of tattooing and scarification or body-modification. However, the currency of the artistic 'mark' particularly since the turn of the millennium aligns with a generational disaffection with the ironic and parodic temper of late-twentieth-century postmodernist culture. In the past two decades, there has been a return of a taste – verging, at times, on piety – for the handmade: from the rough and ready-at-hand to the virtuoso if banal hand-crafted object, demonstrated for instance in Tom Friedman's meticulous pencil-shaving spiraling in a continuous barely visible strand from ceiling to floor, as well as the noisily didactic, sprawling 'temporary monuments' of Thomas Hirschorn, made from packing tape, cardboard boxes and photocopies.

This shift from the ironic museological idolatry of ready-made objects (such as Jeff Koons's pristine vacuum cleaners in vitrines or Damien Hirst's stuffed shark) to pragmatic ready-made material is allied to an emergent discrimination for unaffected directness dubbed the 'new sincerity'. This is a term that tends to invoke hyperbolically honest, candid, amateur and therapeutically communitarian enthusiasms such as flourish in reality TV melodramatics, on Facebook and in the blogosphere. But at its other extreme, the imperative for direct speech and for open source can drive aggressive interference with the globalized digital economy – invoking piracy, hacking, cracking, ripping and the unedited decanting of government and corporate confidential documents.

Thomas's allusion to the mark as the germinal motif for his nanotech aesthetics has correspondences with this new cultural milieu. And we understand this mark, in Thomas's terms, as a sort of folly of nanotechnology. Thomas perceives the 'facture' of the artist's mark-making not in terms of 'manufacture' (made by hand), but as 'nanofacture': as a mode of handling of material at the level where boundary distinctions between medium and agency dissolve in uncertainty. Of course, 'handling' is as anachronistic here as the terminology of 'cutting and pasting' is for word processors. For Thomas, facture is the transference of energy states that occurs at a fundamental level of being, a germinal level, marked by 'touch'. This touch reverberates in numerous contexts of usage: it is proximate or tangential physical contact but also an empathic sentiment; it is an interference, imposition and even molestation but also the slightest degree of tinting or tainting; it is an effable characteristic or mannerism but also a sign of inspired madness; it is transgression but also bonding.

The artistic manifestation of this 'nanofacture' was crystallized in Thomas's *Midas* project from 2006, which focused on the transference of gold atoms and the atoms of human skin cells when they are brought into contact. But in this case, Midas's mythological touch – that turned whatever he came in contact with into gold and, by literalizing his greed, revealed the divine gift as a curse – is something like Feynman's pantograph: it is an analogy that must be pushed across a quantum leap by a sort of symptomatic slip. Thomas's work does not illustrate the moral of the mythological story, even in an ironical way. Indeed, his work might have been called, after Ian Fleming's supreme villain, *Goldfinger*: unlike Midas, whose poetic *anagnorisis* reveals the true object of desire as the thing whose form he cannot ever touch (the food he wishes to eat, the body he wishes to embrace), Goldfinger knows that his desire is precisely the incorporation of a substance that can take any form he wishes it to, from ingots to the lethal gold paint on a woman's skin. In Thomas's work the finger that touches the gold is touched by the gold, but not in any simplistic reciprocation or symmetry. The artist's gesture is moved by its medium, infatuated like Pygmalion with its own touch, moved in an affective register like Goldfinger, and also moved literally at an atomic level. It is not that the artist's fingertip fuses with the gold as a metaphor for some sort of deep connection between artist and medium, but that both are 'swerved' in their asymmetric desire to touch and be touched.

The mythological Midas provides a moral lesson for artists, similar to that offered by another great tableau of the infectious influence of gold: the Golden Calf of Aaron, from the Jewish Testament. The idol of the Golden Calf is any artist's dark mirror, since it is the embodiment of the luxurious and libidinal jeopardy occasioned by love of 'graven images' and aesthetic sensation. In a way, Midas provides a sanction on this danger – or rather the true object of his desire does so, in the words of the resurrected Christ: *noli me tangere*, do not touch me. This museological sanction however is transgressed by Thomas's 'nanofacture', but infringed through a slip or swerve rather than a violation. The facture of Nanoart will be rendered through the over-extension of the apparatus, that 'pantograph', which provides us with means for visualizing that which cannot be touched, and the aesthetic form of that which is beyond sensation.

**Edward Colless**

## Notes

1 *Engineering and Science*, California Institute of Technology, 23: 5, February 1960, p. 3. Feynman's talk was delivered during the American Physics Society's annual meeting, on 29 December. The adapted text of his paper is printed on pages 22–36 of that issue. Original magazine layout is available at http://calteches.library.caltech.edu/47/3/ES.23.5.1960.0.pdf. Retrieved 15 June 2012.

2 Of the 58 pages in that issue of *Engineering and Science*, 35 are full-page advertorial. Feynman's paper is interspersed with advertisements for Western Electric, Pratt & Whitney aeronautical engineering, Urban Carbide, Grinnel engineering, Lockheed, RCA, Sylvania Electronics, a drop forging factory, and Alcoa Aluminum.

3 Consensus opinion accords K. Eric Drexler's 1986 publication *Engines of Creation* that honour: a publication self-evidently more influential in defining the directions of molecular nanotechnology, allowing for its summative and synoptic treatment of scientific and industrial research unavailable to Feynman 26 years earlier.

4 'When I make my first set of slave "hands" at one-fourth scale, I am going to make ten sets. I make ten sets of "hands", and I wire them to my original levers so they each do exactly the same thing at the same time in parallel. Now when I am making my devices one-quarter again as small, I let each one manufacture ten copies, so I would have a hundred "hands" at $1/16^{th}$ size. Where am I going to put the million lathes that I am going to have? Why there is nothing to it … If I made a billion little lathes, each 1/4000 of the scale of a regular lathe … there is less than 2 per cent of the materials in one big lathe' (Feynman 1960: 34).

5 Feynman actually invoked this old Scholastic, medieval conundrum by offering a challenge to reduce the contents of the *Encyclopaedia Brittanica* etched into the metal head of a pin (Feynman 1960: 22). The award (the second 'Feynman Prize') was given in 1985 to a graduate student who succeeded in reducing the first paragraph of *A Tale of Two Cities* by Feynman's required pinhead ratio of 1:25,000.

# Introduction

It is a physics, and, in a given flow, the clinamen is experienced, required by experimentation.

(Serres 1982: 102)

To begin this publication I want to make clear that my intentions are not to write a scholarly work about Nanoart, but to explore ideas, thoughts and concerns from my experience as an artist. The book will reference and research key areas relevant to Nanoart through a process of historical experimentalism. The content for this book has been based on a self-emergent process that explores an art historical understanding of matter and uses various touchstones to elucidate the effects on materiality and agency as a result of the emergence of nanotechnology. The blurring of material boundaries will be reflected in my establishing a fluid organic spatial narrative in which to place ideas, propositions and concerns.

The emergence of Nanoart, through this publication, will be contextualized in relation to specific scientific instrumentation and discoveries. The publication will reflect on the practice of selected artists currently working in the area of nanotechnology and position their works as touchstones for exploring our understanding of materiality from an atomistic viewpoint. This book does not intend to refute the science of nanotechnology but to investigate how this field of research affects, influences and enhances artistic theory and practice.

Why write a book on Nanoart? The desire stems from my own artistic practice, which seeks to question and reconfigure our human understanding of being in the world. In this endeavour, the role, use and creation of scientific, machinic, instrumentalized visualizations will be explored and challenged in connection to specific artistic research. I am working from the concept that science is not an absolute truth, but a quest for the best interpretation of a hypothesis or theory at any given moment. The rhetoric surrounding nanotechnology states that it can rebuild nature 'atom by atom' and has the potential to feed our medical and materialistic intentions. With the ability to construct matter from bottom-up technologies, there is the intention of being able to manufacture organic and non-organic materialities. This approach will test our understanding of what it means to be human and raise new questions about human beings' boundaries.

What is Nanoart? There have been a number of descriptions of what signifies Nanoart. One definition of nanotechnology is based on working with matter whose dimensions and tolerances are less that 100 nanometres (1 nanometre = 1 billionth of a metre). When confronted by this sense of scale it is hard to fathom in our minds what this space could actually look like. As a guide we only have our own sense of scale and size and that has, in part, been controlled since the Renaissance by a perspectival and stereotypical portrayal of ourselves as the centre of the universe.

Nano is Greek for dwarf and the word nanotechnology 'was first proposed in the early seventies by a Japanese engineer, Norio Taniguchi, implying a new technology that went beyond controlling materials and engineering on the micrometer scale that dominated the 20th Century' (Gimzewski and Vesna 2003).

Stefano Raimondi in his catalogue essay states that 'Nanoart is a creative, aesthetic process, which makes use, both in its research and its realisation, of nanotechnology' (Raimondi 2007: 121). We live in an era of nanotechnology and what is being revealed by scientific discoveries and prophesies is shifting our subconscious understanding of the material world. Contemporary art of the twentieth century had been at the forefront of experimenting at the intersection of science and culture to present a human experience and critique on what is being revealed.

A cornerstone of the book is the concept posed in the philosophical writings of Lucretius of the unpredictability of the atoms' swerve and its formative role in the beginning of all matter, form, life and individuality. The connection between atoms, the swerve and individualism will be related to specific artists' pursuits and will involve looking at the works of, amongst others, Umberto Boccioni, Frederick De Wilde, Laurent Mignonneau Claude Monet, Mike Phillips, Georges Seurat, Kevin Raxworthy, Christa Sommerer, Paul Thomas and Victoria Vesna. Artists working with nanotechnologies rarely work alone and therefore this book will also explore the relationship between disciplines.

The information given to us by the Roman Epicurean Lucretius is explored via the philosophical work of Henri Bergson, Giles Deleuze, Felix Guattari, Michel Serres and Gilbert Simondon who posit a way of thinking that is linked to the unpredictability of the swerve and compare it with a life that has no swerve.

The reawakening of material significance in a post-ocularcentric world is in the process of becoming through the haptic regime of the technological interface. These cumulative influences of becoming are now confronting not only ocularcentrism but also humanocentrism.

One of the main ongoing touchstones will be presented in Chapter 1. This chapter sets out the contextual relationship between art, in particular drawing, and the materiality of carbon. This is a concept that will be developed throughout the following chapters. Our understanding of carbon's material structure at an atomic level has undergone a meteoric rise over the last 25 years as its complexity is being revealed. Carbon is one of the oldest known materials and is the fifteenth most abundant element in the earth's crust; the fourth in the universe after hydrogen, helium and oxygen. Its presence in all known life forms allows for this substance to be a catalyst in reconfiguring a humanocentric understanding of the material world.

Two very well-known forms of carbon are graphite (from the Greek word *graphein* meaning' to write) and diamonds (from *dámas* meaning unbreakable). Other structures that have been discovered relative to carbon atoms include strings of atoms in 1970, cages of atoms in 1985, nanotubes in 1991. Graphite is formed from atoms that are held together in a hexagon shape that can be easily reduced to a single sheet of atoms, which is called graphene. In 2010 the physics Nobel Prize was awarded to Andre Geim and Konstantin Novoselov for their experimental research with graphene.

A main criterion of linking carbon and other forms of matter to the expressive world of art is through drawing. Drawing with its use of graphite will be explored in connection to the initial marks of inquiry made by the artist that were made to understand and represent the world. The use of graphite in pencils is explored through the development of our contemporary understanding of carbon, which has been influenced by science. The importance of drawings in representing and visualizing our thoughts will be discussed in context to early experiments in the use of materials like silver in silverpoint drawings. Silver also has a significant importance in our construction of visualizations of the world using mirrors, photography and film.

Drawing is a direct connection with what is being seen, sensed or felt. It allows for marks to be made that position thought in the metaphorical infinite space of the sheet of paper. The process of making marks positions us within the world and can be seen as an ongoing set of approximations. Each mark supports, interacts and justifies each other's place on the surface of the white void. Matter's interrelationships and connections are brought into existence through what Paul Klee suggests is point, line and plane. The artist's remit to explore the material and immaterial world can be seen through the process of plotting the world through drawing.

Renaissance perspective alienated us from the material world with its construction of a godlike vision. All matter in the world post-perspective was available for the individual's ownership via the gaze. At this point in time a rupture in the way we saw the world happened so that we no longer viewed the material world as a space in which we existed as a totality but more as something we observed. The potential for ownership and control of resources empowered a new era of environmental domination and vandalism of the material world to occur.

Chapter 2 looks at the main instrument of nanotechnology, the Atomic Force Microscope (AFM) as it has evolved to reveal atoms using the touch of a cantilever.

The AFM uses a cantilever with a pointed tip of 10 nanometres to record the specimen by various contact modes that all involve touch. This aspect of touch changes the way that we think about imaging the world, it no longer privileges sight. This chapter explores the software that interprets AFM data and allows the atomic structure of the world to be revealed, imaged and manipulated. The AFM has the ability to move atoms and creates opportunities for bottom-up technologies to develop a mantra of rebuilding the world atom by atom. The AFM will be put into the context of artists and scientists who have demonstrated its use and demystified its mechanics.

The discoveries that have been driven by the invention of the AFM will be contextualized in relation to an art historical understanding of the machinic image. Parallels will be drawn between a history of making the invisible visible in art at the turn of the twentieth century, the influential work of Etienne-Jules Marey and Eadweard J. Muybridge in illustrating the previously unseeable motion of the human body, and the imaging created by the AFM today. Marey's and Myubridge's photographic experiments were highly influential in visualizing and analysing movement on artists working at the turn of the twentieth century. The evolutionary development of the microscope from optical to AFMs will be explored in relation to an art historical context of making the invisible world of atoms visible via touch. This invisible world of atoms is now being revealed through scientific instrumentation and software that interprets the data. The immaterial substrate of the universe is therefore being revealed, visualized and made malleable.

The influential work of Marey will be the main focus in linking scientific photographic research in relation to movement with the philosophical theories relating to duration and matter developed by the philosopher Henri Bergson. Bergson's understanding of duration and matter can be seen to underpin the work of a number of artistic endeavours that attempt to look beyond the picture plane and into the world of atomism. Media artists practising since the seventies have innately critiqued the human relationship to technologies and the implied immateriality of technological practice. This fortuitous relationship has maintained an ongoing interest in what constitutes materiality in the arts. The artworks of Boccioni, Mignonneau, Sommerer, Raxworthy and Thomas are used to demonstrate an artistic awareness in the significant shift that is taking place towards an acknowledgement of materiality as a result of nanotechnological recording of atoms through touch not sight.

Chapter 3 focusses on the concepts of vibration, vitalism, life and materiality and extends the artist's concepts of agency in relation to matter. The purpose is to explore how the artist's practice can be linked conceptually and contextually to examining, exploring and experiencing the atomic world. This chapter examines an understanding of how the world becomes de-territorialized through the discoveries of nanotechnological research and how immateriality of matter extends our understanding of culturally defined material agency[1]. An exploration of the works *Blue Morph* and *Nanoessence* will be key in examining what constitutes living in a nanotechnological world.

From the primordial swerve the role of vibration, rhythm and pattern have been cited as being at the very core of the evolution of matter. This will be explored in context of the work of artists such as Ann Niemetz and scientist Andrew Pelling. This chapter explores the role of the swerve and the concepts of unpredictability with regard to humanities' changing understanding of what constitutes vitalism or the life force. The context of unpredictability will be explored through the experimental artistic and scientific practices that explore vibration and transmission.

Chapter 4 looks at future visions for reconfiguring our understanding of matter and what artists are adding to the discourse. The mechanistic interpretation of the material world into quantifiable data is seen and challenged. It examines the work of artists who explore scientific measurement in order to transcend the role and rhetoric of nanotechnology and create new forms of social engagement with matter.

The artist Frederick De Wilde's work *Hostage* will be explored in the context of revealing how new meaning can arise from nano scientific research into nanotubes and blackness. The work *Mote* by Mike Philips will be highlighted as it uses a speck of dust in becoming a mediator between the digital and analogue. Raxworthy and Thomas's recent work *Atomism* re-examines the topic of the mirror that Feynman refers to specifically in his 1979 lectures in New Zealand (Feynman 1979). How photons operate in a mirror is contextualized with regards to an art historical look at the mirrored surface. It raises questions about what is revealed and what is lost or seen and not seen.

The book looks at artists' works that draw metaphorical pictures to hopefully encourage an intellectual and cultural critique on our understanding and relationship to the material world. The question of whether we can change our relationship to non-organic matter and its temporality is to be revisited. The relationship to matter has to be non-perspectivalized, not hierarchical and not valued by humanistic illusion of comparison. The domination of the world as a material resource to be violated and extracted as fuel suggests a blindness to the evolution of a non-materiality.

## Note

1 Material agency in the context of this book is related to the way that matter can mediate and transform the way we humans interact with the world. The idea of material agency is in part changing the embedded nature of humans towards the material world as the properties and meanings of matter evolve.

# Chapter 1

## Materiality and Immateriality of Art in the Age of Nanotechnology

All pictorial form begins with the point that sets itself in motion … The point moves … and the line comes into being – the first dimension. If the line shifts to form a plane, we obtain a two-dimensional element. In the movement from plane to spaces, the clash of planes gives rise to body (three-dimensional) … A summary of the kinetic energies which move the point into a line, the line into a plane, and the plane into a spatial dimension.

(Klee 1961)

The book consists of a series of reflections that draws on metaphor, abstract analogy, personal associations, connections and critical reflections that are more poetic than based on concepts of fact. I will begin this journey by exploring a seminal tool in art by looking at the humble pencil via its materiality and history and by drawing on analogous relationships to atomism and nanotechnology. The pencil is a medium that as an artist I have used for most of my working life. As a tool it has been instrumental in my continual reassessment of the world around me. The pencil enables me to draw out meaning through the very act of mark making, and by extension, through the analysis of drawings and the dissemination of the act of drawing through teaching. My aim is to reassess the role an artist has with the pencil, and the agency[1] and meaning inherent within the marks made in the wake of emerging technologies and scientific discoveries in the area of Nanoart.[2] I will be focussing on the analogous relationship between the drawing pencil, carbon and current thinking in the area of quantum computing.

Terms such as 'nano' are now embedded within our understanding of the world, creating a homogenization between science and culture. The evolving research of nano materials, concepts and processes challenge our understandings of boundaries, the body and the material object world. Nanotechnology[3] has created new understandings of materials and processes and is subsequently shifting the social understanding of our own humanity. Developments in nanotechnologies seriously impact on our understanding of materiality.

Two terms often used interdependently are nanoscience and nanotechnology. Surprisingly, the term nanotechnology predates nanoscience. This is because the

> dreams of a new technology were proposed before the actual scientific research specifically aimed at producing the technology existed. The term nanotechnology, in its short lifetime, has attracted a variety of interpretation, and there is little agreement, even among those who are engaged in it, as to what it actually is.
>
> (Gimzewski and Vesna 2003: 2)

New materialities created as a result of scientific discoveries have been historically and culturally slowly assimilated into traditional visual art rhetoric and practices. These artistic practices have been historically grounded in traditional materials and techné. However, the inclusion of nanotechnology into this arena has raised questions about the intrinsic nature of matter as an immaterial substrate of art. This chapter begins to explore some of these questions by focusing on the materiality of drawing and its link with the immateriality of the element carbon.[4]

The contextual relationship between theory and practice in Nanoart will be explored by examining its expanding syncretic social relationship to immateriality and materiality. This chapter draws on the Nanoart work of Victoria Vesna and James Gimzewski and their project *Zero@wavefunction* as a seminal work. *Zero@wavefunction* acts as a demonstration of a change in the perpetual and physical experience and expands our understanding and critical analysis of art and nanotechnology.

The links being drawn together throughout this chapter will show how the artist's experimentation via serendipitous relationships is shifting understanding of materiality, agency, space, scale and autonomy.

## Individual atom

A starting point with which to explore our shifting understanding of materiality is by defining the differences between what is considered quantifiable and qualitative in comprehending matter. For example, Titus Lucretius Carus (Lucretius), a Roman philosopher whose philosophical poem *De rerum natura (On the nature of things),* proposed the clinamen as a name for the unpredictable nature of how atoms relate and connect with other atoms.[5] The clinamen[6] describes how atoms falling in a void swerve and make connections. In so doing the atoms create matter and form. He proposes:

> The atoms, as their own weight bears them down
> Plumb through the void, at scarce determined times,
> In scarce determined places, from their course
> Decline a little – call it, so to speak,

Mere changed trend. For were it not their wont
Thuswise to swerve, down would they fall, each one,
Like drops of rain, through the unbottomed void;
And then collisions ne'er could be nor blows
Among the primal elements; and thus
Nature would never have created aught

(Lucretius Carus [c. 50 BCE]: 1950)

In this extract, Lucretius defines the poetic entanglement of matter and how it is inscribed with characteristics based on the random nature and unpredictability of the swerve.[7] These primordial connections, according to Lucretius, are the basis for the production of nature. The free-falling atoms in the void are not prone only to the laws of the fall that occurs endlessly in space. If it were not for the smallest unpredictable turbulence causing the atom to swerve and produce the first atomic forms, then there would be nothing. The swerve is the first evolutionary step towards something other than what was; it is the beginning of nature. This concept of chance association as a form of nature arises when we consider forms in which each atom, being a unique link within the structure, creates alliances based on the randomness of the swerve. This concept of chance association in nature presents as uniqueness.

Gilbert Simondon builds on the concept of nature's uniqueness, exploring it in relation to individuality.[8] He proposes that the term 'individual', pre-dates the individual and argues that for a thing to be individualized it needs to be based on a specific existence. He states:

> The atom interacts with other atoms through the *clinamen,* and in this way it can constitute an individual (though not always a viable one) across the entire expanse of the void and the whole of endless becoming. Matter can be impressed with a form, and the source of ontogenesis can be derived from this matter-form relation. Indeed, if haecceities were not somehow inherent within the atom, or matter, or indeed form, it would be impossible to find a principle of individuation in any of the above-mentioned realities.
>
> (Simondon 1992: 298)

> Those cohesive forces themselves, which may be taken as the principle of individuation of the complex individual, are in fact negated by the finer structure of the eternal elementary particles, which are the real individuals here. For atomism, the principle of individuation is rooted in the very existence of an infinity of atoms; it is always already there as soon as thought seeks to grasp their essential nature.

> Individuation is a fact: for each atom it is its already given nature, and for the complex unit it is the fact that it is what it is by virtue of a chance association.
>
> (Simondon 1992: 299)

Thus, while randomness creates form, which in itself is individualistic, it also becomes a seminal concept of individuality. The individuality of the body then exists embedded within the atomic structure of the body. The atoms that are inscribed with a 'given nature' are in turn making connections with other atoms, matter and form.

The start of our individuation runs parallel to all of nature where unpredictability is bestowed on all matter. Therefore it makes a case for its own moments of evolution. Michel Serres, when talking about the clinamen, refers to the ongoing metaphorical struggle between Mars and Venus. He expands upon Lucretius' *On the nature of things* and suggests:

> [T]he law is the plague; the reason is the fall; the repeated cause is death; the repetitive is redundance. Everything falls to zero ... The angle of inclination cures the plague, breaks the chain of violence, interrupts the reign of the same, invents the new reason and the new law ... gives birth to nature as it really is. The minimal angle of turbulence produces the first spirals here and there ... Turbulence perturbs the chain, troubling the flow of the identical as Venus had troubled Mars.
>
> (Serres 1982: 102)

The repetition inherent in the act of falling is exemplified in the digital rain during the opening credits of the motion picture *The Matrix*[9] It may be seen as a visualization of Serres' statement that 'repetitive is redundance'. The world of the fall is the problem to which the clinamen brings the swerve. The swerve offers up an understanding of the world of 'nature as it really is', of its uniqueness and individuality. Serres states: 'Atoms are not souls; the soul itself is atomic' (Serres 1982: 103).

The unpredictability of the swerve as offered by Lucretius, the individuation through atomism presented by Simondon and the atomic soul of Serres are keys to the conceptual construction of the experimental relationship between carbon, the artist's pencil and thought.

## Drawing the future

The drawing pencil as a tool for the visualization, ideation, and for creating plans, diagrams, doodles and representation of the world is well documented. The process of drawing is creatively thinking on paper. It is not merely to draw out ideas or to represent things, but it helps us to see, comprehend and to sense more holistically. The art of drawing is a discursive act in which the pencil as a tool is mediating between the seer and the seen. The practice of drawing enables the artist to rethink the marks being applied to the surface in terms of ideas to be reflected upon and presented back into the world.

The traditional drawing pencil[10] invented in 1564, was developed after a large deposit of graphite was found in the United Kingdom. Graphite is derived from the Greek word *graphein*, meaning to draw or write. The graphite needed to be encapsulated in a wooden tube, because its structure was more brittle in nature than its lead predecessor. If we examine the graphite material inside a pencil at an atomic level, it appears as a series of carbon atom sheets, each one-atom thick (Figure 1).

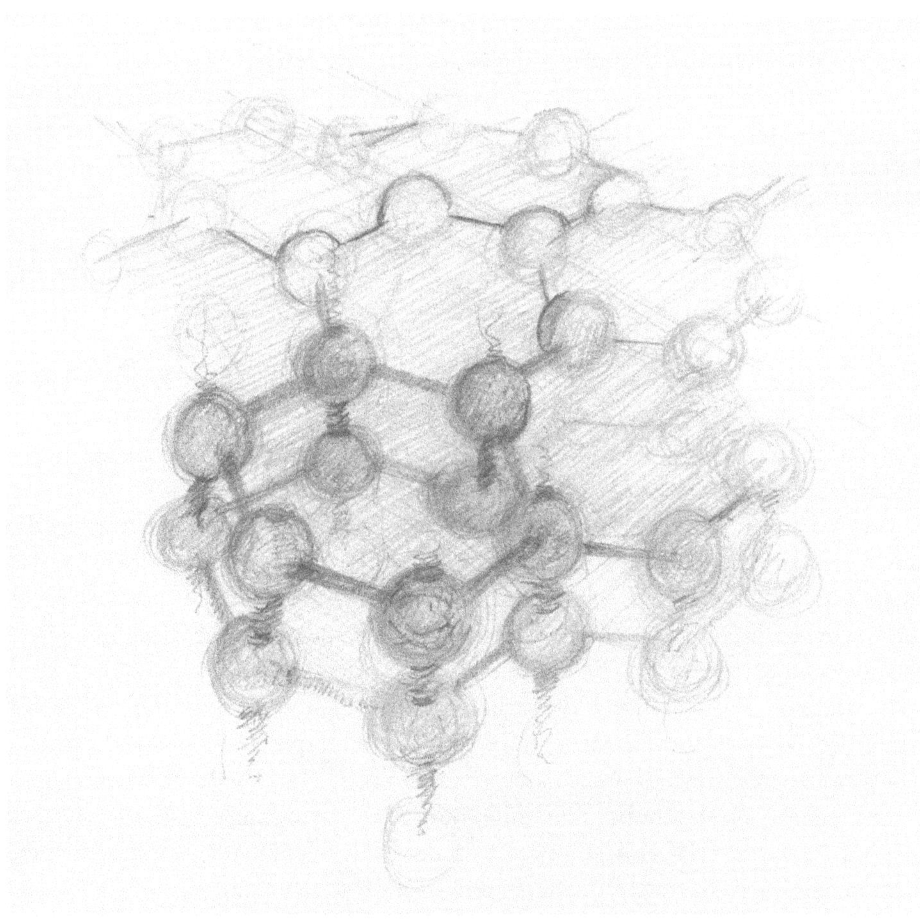

**Figure 1:** Drawing showing the layers in sheets of graphene (carbon atoms). By Paul Thomas (2010). Courtesy of the artist.

The sheets of atoms called graphene which make up graphite, are weakly bonded on top of one another by the van der Waals force.[11] This weak bond causes each sheet of graphene to be unstable and allows them to be easily transferable to surfaces such as paper. The artist uses touch to apply varying deposits of carbon to the paper. The process of drawing means that marks made by the artist are indicative and intentional. Thus the artist is not only applying graphene sheets to a surface, but also thinking through the pencil and potentially imbuing the sheets with thoughts.

While we are familiar with the use, agency and aesthetics of graphite in drawing, the understanding of the materiality of graphite is changing and therefore new agencies and aesthetics of graphite are becoming part of our conscious understanding of the world. The next section will examine the current uses of graphite, in particular in the construction of the quantum computer. I will examine how making comparisons between the use of graphite as a pencil and the extended use of the material, for example, in quantum computing, potentially shifts the perceptual context of the material substance. This presents opportunities to rethink and recontextualize organic and inorganic matter, and this material form's intrinsic relationship to life.

## Buckyballs and nanotubes

The known family of carbon elements grew in 1985 when Sir Harry Kroto, Richard Smalley and Robert Curl discovered the Buckminsterfullerene carbon molecule $C_{60}$. The 'buckyball' as it came to be known, was named after Richard Buckminster Fuller who developed architectural geodesic domes that resembled the carbon molecule $C_{60.}$ The buckyball is constructed from the same atomic structure as the sheets of carbon atoms that make up graphite, but the atoms form a ball shape similar to the pattern of a modern soccer ball.[12] As will be discussed shortly, the hollow carbon buckyball has played an important role in the development of the quantum computer.

In addition to the sheets of carbon atoms (Figure 1) and the buckyball, nanotechnology research has more recently discovered graphite, as found in the pencil, which can be split down to single sheets of carbon atoms (graphene[13]). The one-atom thick graphene sheets can be rolled up and used to create single-walled nanotubes (SWNT). Each SWNT is a seamless cylinder with a diameter of approximately 1 nanometre (approximately 10 atoms) and a longer tube length. Nanotubes exhibit extraordinary strength and so are currently being tested, amongst other possibilities, in the construction of quantum computers.

## The quantum computer

Quantum computing is based on properties of quantum mechanical phenomena such as superposition. Simon Benjamin points out that 'individual atoms can support quantum superpositions for long periods, and such atoms can in principle be embedded in a permanent molecular scaffolding to form an array' (Benjamin et al. 2006). Thus, one possible building block of the quantum computer is the buckyball $C_{60}$. The $C_{60}$ molecules have individual nitrogen atoms fired at them that become trapped inside the buckyball, creating the $nC_{60}$ molecule, which is at the very core of a microprocessor's array. The caged nitrogen atom within the buckyball is incarcerated and isolated but still has the ability to spin and can be controlled by microwaves to emulate the zeros (by spinning up) and ones (by spinning down) of the digital computer's binary code. However, quantum computing explores the potential of the nitrogen atom's superposition inside the buckyball to be in more than one state (not zero or one) at the same time.

This theory of being simultaneously in more than one state in relation to quantum mechanics can be explained using the Schrödinger's cat thought experiment. In his original thought experiment of 1935, Erwin Schrödinger imagined that a cat is locked in a box along with a Geiger counter and a radioactive source connected to a vial containing a deadly poison. If the atom decays and the radioactivity is detected by the Gieger counter, it causes the vial to smash and the cat to be killed. However, when the box is closed, we do not know if the vial has smashed or not. The cat can either be in an alive state or in a dead state. We might understand the cat to be alive and dead at the same time, a state explored in explaining quantum physics.

A framework for the construction of the quantum computer, using buckyballs, requires the $nC_{60}$ molecules to be positioned in an array that can receive and transmit information. So the buckyballs, due to the natural attractions of the van der Waals force (as discussed previously), can be drawn into the SWNT. Here the buckyballs' nitrogen atoms' isolated atomic properties, incarcerated in the buckyball and trapped in the Nanotubes carbon tube, can be placed in an array over a series of connectors on a substrate, to form the basis of the data processor in quantum computing. These arrays become the platform on which information can be processed and ideas can become synthesized. In effect, atoms become the tools to enable information to be processed.

The buckyball with the nitrogen atom inside, $nC_{60}$, is analogous to the Schrödinger's cat thought experiment. The $nC_{60}$ entrapped within the nanotube is given sufficient impulse from a microwave; it can make the nitrogen atom spin to face up, north, alive, or face down, south, dead. When a shorter microwave pulse (10 nanoseconds) is applied, it spins the nitrogen atom to a superposition somewhere between up and down, north and south, alive and dead. When the same microwave pulse is applied to the nitrogen atom, the atom will return from a superposition somewhere unknown back to its starting point. Using this superposition potentially allows more data to be attached to the spin of the atom than a conventional computer and thus the $nC_{60}$ can be used to create two superpositions that allow the processing of more data than a conventional computer (Benjamin et al. 2006).

Nanotechnology and quantum computing are exploring shifts that challenge our notion of the machine. The 'clinamen' concept, if applied to each atom, is analogous to the emergence of the quantum computer. The swerve or spin is creating more than just information, it is the probability of these data existing in a parallel state.

## The Atomic Force Microscope

A fundamental change in our conscious understanding of materiality has evolved through the adoption of a quantifiable 'machinic' understanding of the world. Furthermore, through the instruments of nanotechnology, a change has occurred in our conscious understanding of materiality. We now see evidence made visible in the imagery produced by instrumental tools of nanotechnology such as the Scanning Tunnelling Microscope (STM), 1981, and later the Atomic Force Microscope (AFM), 1986. The AFM is one of the foremost tools for imaging, measuring and manipulating matter.

Optical microscopes were used exclusively until 1931 when the first electron microscope was introduced. Optical microscopes have a number of limitations in that they can only image dark or strongly refracting objects effectively. The diffraction limits resolution to approximately 0.2 micrometre and ambient light can diffuse the focus. Unlike the previous microscopes, the AFM is not optical. It is one in a general category of Scanning Probe Microscopes (SPMs) that includes both the STM and the AFM and its name, in fact, is a misnomer. The AFM uses a cantilever probe with a pyramid-shaped tip measuring approximately 10 nanometres. The tip touches surfaces, in a scanning process. An image of the surface is obtained by mechanically moving the probe in a raster scan of the specimen, line by line, and recording the probe-surface interaction via a laser reflected on a photodiode.

The concept of gathering scientific data through touch allows for a reconfiguration of a dominant ocularcentric understanding of the world. The AFM and nano assembly show the potentiality for moving atoms around to create new materials, forms and structures thereby replicating the construction of humans from dust particles.

## Nanoart

The new materials like graphene that are brought to us by scientific research can be explored, critiqued and contextualized by artists who have traditionally used graphite and now graphene, to change our understanding of a material's agency. We can no longer see the material of graphite without the contextual relationship with its use in quantum computing and electronics. These contextual relationships are confronted via Nanoart to visualize and change the way we think and experience the world of atoms.

I will examine the work *Zero@wavefunction* produced collaboratively by Victoria Vesna and Jim Gimzewski, and support from Josh Nimoy, Pete Conolly and David Votava which premiered at the Biennale of Electronic Art Perth in 2002. Vesna and Gimzewski *Zero@wavefunction* project is a significant Nanoart installation using a carbon molecule $C_{60}$ buckyball as its central motif.

In the installation, images of machine-generated buckyballs were data-projected onto a large wall in the John Curtin Gallery. Viewers within the gallery used their larger-than-life shadows, which is the absence of light and not material objects, as a prosthetic device to manipulate and interact with the buckyballs' shapes and movements. The interactive artwork allowed users to explore the carbon molecule $C_{60}$ through the immateriality of their own shadow, creating an embodied experience of the gesture. The installation enabled viewers to connect with ways in which nanoscientists manipulate individual molecules billions of times smaller than commonly experienced in day-to-day life.

Interaction with the atomic world, now made visible, is facilitated through the absence of light caused by the immaterial self. The now-visible molecules can be shaped and bent, reforming the substrate of the universe and alluding to the warping of quantum space.

Reflecting on the evolution of art and nanotechnology, Vesna and Gimzewski have spent the last ten years exploring, '[what] will make a fundamental shift in our conscious and unconscious minds'. They suggest:

> As the perception of reality shifts to the collective level, we will find ourselves in an entirely new world, with very different values and motivations. However, we do acknowledge that any radical proposition, with such enormous and global implications, will undoubtedly have to face fierce opposition from those who have so much invested in the old, mechanistic, world-view.
>
> (Gimzewski and Vesna 2003: 12)

The fundamental shift that needs to take place is historically linked to a number of seminal moments when scientific research developed new understandings of our relationship to the world. Walter Benjamin in *The Work of Art in Age of Mechanical Reproduction,* responding to the question of whether photography was art, said: '[the] primary question – whether the very invention of photography had not transformed the entire nature of art – was not raised' (Benjamin 1969). Similarly, the primary questions in relation to more recent technological developments also face 'fierce opposition' from fine art disciplines. The pertinent question now being whether the AFM, and similar technologies, have once again transformed the material nature of art.

This challenge to our understanding of materiality is played out in Vesna's and Gimzewski's artwork, which enabled users to draw and shape molecules with their own shadows (figure 2), thereby demonstrating the development of immaterial relationships while engaging in the energetic possibilities of particles and the world of atoms. The ability to create a credible and tangible interface with the atomic world, by allowing the scale of a user to grow and fall based on their position in

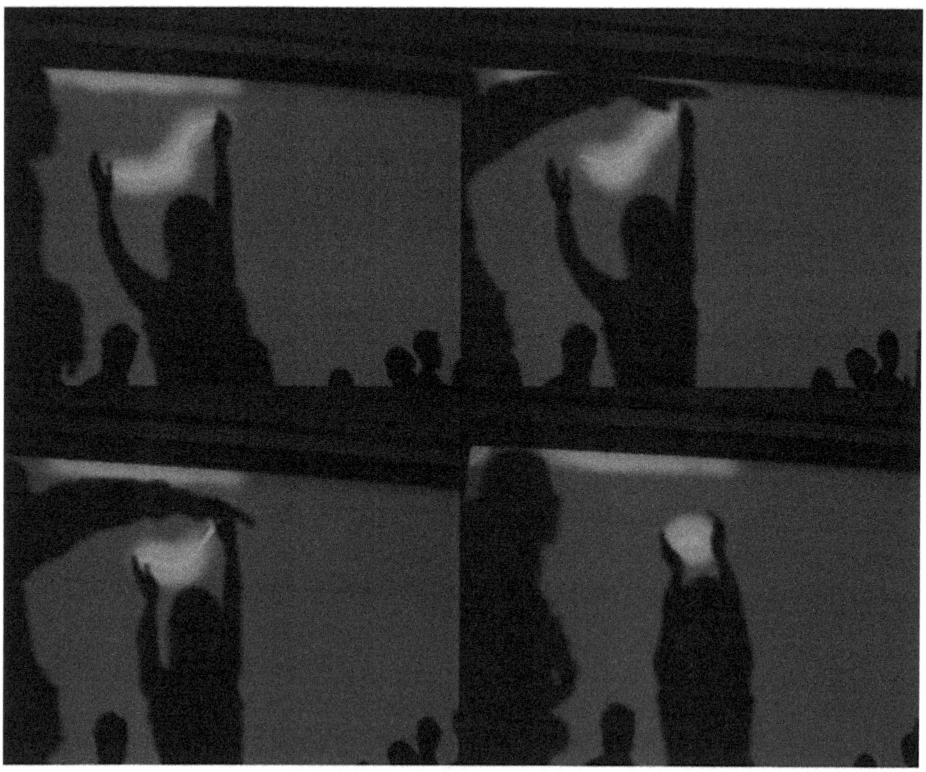

**Figure 2:** Victoria Vesna and Jim Gimzewski, *Zero@wavefunction*, (2002). Stills from video of installation at the Biennale of Electronic Art Perth, Perth. Courtesy of the artist.

relation to the light from the projector, mimics forms of magnification. The user inadvertently draws the buckyball across the surface and engages with it through the gesture in the act of mark making.

The interaction with the work forms complex relationships with matter emulating a playful control by the viewer, but this also opens up the questions of manipulation and colonization of the material world.

## Nanotechnology machinic vision

The development and commercialization of nanotechnology has reinforced a machinic perception of the world similar to that Henri Bergson (1859–1941) critiqued at the turn of the twentieth century. The data images created by the main tool of nanotechnology, the AFM, are constructed from software packages written by computer scientists. The needs of the scientists are not to create beautiful graphic images of matter, but to allow for proximity and measurement to occur. While the technology creates 'machinic' images of matter, in the case of the AFM the images are often underwhelming when compared with the results of an electron microscope.

The AFM imagery is created by a point-cloud system. Colours are selected by scientists from a series of options to represent height maps and create visual references. It is these references that are presented, generated by new codecs and assimilated. The images become part of 'pre-digested' human perception, that is, they conform to the ways in which we might already see the world.

The images are apparitions, they form a haptic visualizing system, created in the same way as a pencil rubbing is created. The data, based on the x, y and z coordinates, create 3D images that can be printed through rapid prototyping to become manifestations of objects. Thus, the physical integrity of any object can now be confused by nanotechnological rhetoric. Bottom-up technologies enable objects to be created that assimilate atom-by-atom ground-up design and the atomic particle structure of an object's chemical make-up. The artefact in this context is reduced through nanotechnology to a mimetic machinic construct, which is held together with orchestrated forces of rhythm, pattern and vibration.

## Science fiction

There has been a significant attitudinal change in recent years among artists, sociologists, philosophers and scientists who see their individual disciplines expanding beyond their 'traditional' domains. This comes at a time when global economics, fuelled by new developments in science and digital technology, provide increasing opportunities for artistic and technological cross-disciplinary interaction.

Colin Milburn focuses on the fundamental changes in our cultural attitude towards nanotechnology. He suggests that change initially grew out of science-fiction narratives where possibilities in relation to scale and matter are played out. The change in understanding brought about through these narratives has opened the door for us to consider evolutionary nanotechnological futures (Milburn 2002). These visions of the future influence cultural and scientific interpretations of matter and materiality, and confront the social role of the artefact. In his cyberpunk novel *The Diamond Age*, Neal Stephenson constructs a contrast between a Victorian attitudinal metaphor and the world of nanotechnology (Stephenson 1995).[14] He draws on a binary opposition that is often based on a fear of nanotechnology brought about by exaggerations made in both science fiction and science.

In referencing the role science fiction played in the development of Richard Feynman's seminal 1959 paper, 'There's Plenty of Room at the Bottom', Milburn demonstrates that fiction has played a large part in envisioning the evolution of nanotechnology (Milburn 2004). In this paper, Feynman makes predictions about the potential for nanotechnology to create fields of investigation such as nanowriting, nanocomputers and nanosurgery. Milburn demonstrates how fiction informs science, and fiction influenced Feynman, by citing Robert Heinlein's 1942 novella, *Waldo*. The hero of the novella creates a mechanical hand which in turn creates a half-size hand, which in turn creates a further half-size hand and so on, in order to enable him to carry out surgery on a sub-cellular level. Milburn's investigations into the source of scientific truth shows that the 'originality of the Feynman myth crumbles, for we can see that his talk emerges from genre science fiction. His method of molecular manipulation is borrowed from Heinlein' (Milburn 2004: 123).

Science fiction becomes factual as myths, which Milburn refers to in his chapter 'Nanotechnology in the Age of Posthuman Engineering: Science Fiction as Science' in the book *Nanoculture*, are reflected in government nanotechnology reports. These reports demonstrate that research in the area of nanotechnology (although unsubstantiated) is both ethical and plausible. It is these questions about ethics and plausibility that Nanoart is addressing. It is playing a significant role in bringing them into the context of the visual arts, creating debate and conjecture in the white cube of the galleries. Ongoing challenges concerning agency and materiality overlay existing understandings of how the art object's agency challenges meaning by reconceptualizing and recontextualizing its very material foundation.

## Drawing matter

The role of the artist, as someone who is involved with non-specific speculative processes of critical inquiry, is needed to create an ongoing challenge and contextualization of scientific understandings and approaches with regard to concepts of materiality and immateriality. Nanotechnology research challenges our relational understanding of matter and the imbued meaning implied in the construction of material artefacts. Through Nanoart, nanotechnologies are thrown into question in totally different ways. Questions arise regarding the permanency of materiality and what constitutes the building blocks of matter; the world's 'immaterial substrates'.

*Zero@wavefunction* created an environment in which the viewer could interface poetically with the carbon molecule. The installation demonstrates how *Waldo*'s concept of the half-size hand, and the potential for the manipulation of the immaterial substrate of the world through scaling, can be experienced.

This poetic control of the molecule in Vesna and Gimzewski's work stands in stark contrast to the 2000 National Science and Technology Council report to the United States Congress titled 'Nanotechnology: Shaping the World Atom by Atom', which seems to encapsulate in its mission, a disregard for the integrity of materiality. In its introduction, the report states that 'the emerging fields of nanoscience and nanoengineering are leading to unprecedented understandings and control over the fundamental building blocks of all physical things' (Amato 1999).

In the same report, Richard Smalley refers to nanotechnology as being 'the builder's final frontier', and Horst Stormer states that nanotechnology has given us 'the ultimate toy box of nature – atoms and molecules. Everything is made from it ... The possibilities to create new things appear limitless'. The basic materiality of everything is seen by Stormer as a playground for things to be re-engineered, reshaped and re-formed into a product to be consumed. Chad Mirkin, director of Northwestern University's International Institute for Nanotechnology in Evanston, Illinois, took the idea a little further; he suggests, 'We're using inspiration from life to create new forms of matter' (Amato 1999). These quotes, all taken from a government report, demonstrate a quantitative and non-sensorial reconfiguration of our position in the world. These reports regard all matter, including life, as part of the 'ultimate toy box of nature', and suggest a continuation and reiteration of the 'Newtonian' concept of power over nature.

The ability to reconstruct matter in completely new forms is not original. However, what is new is the means of unpacking living matter and the potential for it to be re-engineered into new forms of life at the nano level. The concept of constructing living matter demonstrates a presumed human sovereignty and ownership of all matter and the right to play in the reconstruction of nature. Alfred Nordmann suggests the challenge for artists is to understand that the 'scientific way of relating to the cosmological image of nanotechnology abandons the claim of a privileged position for human beings in a divine and externally fixed order' (Nordmann 2004: 50). A shift in human perception is needed for a reconfiguration of our position in the world in light of current understanding of matter brought about through research in nanotechnologies.

Drexler gave us a glimpse of what would be possible with 'biological evolution' (Drexler 2007: 80) through the concepts of nano assembly where robot arms would move molecules around to form parts of a complex object. These parts would then, in turn, be compounded to make more parts of parts until the object would be formed.

The argument being put forward suggests that the fixed order of matter is no longer permanent and can be constructed by nano assemblers.[15] There is further shift in our comprehension of what is possible for human beings already separated from an objectified material world of a 'divine and externally fixed order'. The perspectival imprint laid over the realm of nanotechnology allows for a Cartesian understanding of nanospace and raises the spectre of ownership.

However, nanotechnology and Nanoart offer alternative points of view. Contemporary arts have continually challenged perspectivalism, constantly attempting to reconfigure our perception of the world. Similarly, nanotechnology demonstrates the potential to break this dominant ocularcentric position due to the development of instruments for imaging the nano-world being based on touching the material substrate. At the nano level seeing becomes touch and vice versa, thus confronting our understanding of the territorialized colonial space represented by Cartesian logic.[16]

In a similar fashion to how the futurists at the turn of the last century revealed military ambitions in relation to technological developments, the desire for research in nanotechnology is driven by the military. The military are attracted by the potential to explore and utilize de-territorializing expectations and possibilities. The spectre of the post-human soldier with the capabilities to digest huge amounts of data and information reflects a problem inherent in the human's inability to process too much information. In contrast to current situations where bottlenecks of data are formed, created by feedback loops, the new post-human soldier offers other possibilities. The post-human soldier needs to become de-territorialized even to the point of telepathically linking into computer systems. Milburn, in his paper[17] 'Nanowarriors: Military Nanotechnology and Comic Books', implies that

> [the soldiers' super] capabilities would stem from biomechanical exoskeletons and musculature actuators, as well as metabolic redesign of the soldier's body

> against shock, trauma, and sleep deprivation. They could include psionic powers like telekinesis, for through a nanowired 'brain-machine interface', a soldier might command peripheral computers, vehicles, and weapons with thoughts alone.
> (Milburn 2005: 83)

The potential presented by Milburn is of nano-fictional scenarios to be developed as a justification for the construction of military materialism. The de-territorialized soldier challenges materiality. Through him our understanding of being human is blurred between a materialist pursuit of nanotechnology to control matter and our understanding of life. The relationship of thought to the world as a 'nanowired' experience extends the concept of the transportation of thought. Thought being transmitted between all matter raises questions with regard to what constitutes thought and how it is recognized. But what appears to be evolving, is that thought is inextricably imbued in atoms.

We begin to gain an understanding of life and its relationship to the material body at an atomic level as having no defining boundaries. Writing on the basis for life at a nano level, Vesna and Gimzewski reflect on how 'Watson and Crick explicitly described DNA in computer terms as the genetic "code", comparing an egg cell to a computer tape' (Gimzewski and Vesna 2003).[18] Simon Penny also states that '[A] basic premise of Artificial Life, at least in the words of one of its major proponents, Christopher Langton, is the possibility of separation of the "informational content" of life from its "material substrate"' (Penny 1997).

The body that has 18 per cent of its chemical compound made up of carbon holds the graphite pencil at the periphery of its traditional physical boundary as a prosthetic device. When we use graphite from the carbon family as a material its inherent energy and agency are employed when drawing our ideas onto paper. At an analogous level, each pencil mark is also embedded with the information of the subject matter and concept being developed. The graphite molecules being attached to the paper molecules are inscribed with inherent data resulting from this plastic process. In the information feedback loop the hand makes the mark, the brain sees the mark and the hand then responds to both the mark on the paper and the source of the idea. The graphite drawing can be understood as a visualization of an interaction between a thought and a representation. The thought is transferred and embedded into the atomic materiality of the drawn marks. The materiality of graphite is imbued with the thought and therefore some of the thought must be imbued with graphite. Deleuze and Guattari explain: '[An] ancient atomism not only multiplied Parmenidean being, it also conceived of Ideas as multiplicities of atoms, atoms being the objective elements of thought' (Deleuze 1968).

In *Zero@wavefunction*, the viewer's shadow draws and manipulates the carbon molecule that is related to graphite. The viewer in the gallery is drawing the buckyball across the space with their shadow; in this process a series of physical relationships develops with the immateriality of their material gesture. The connection being made to the world of nanotechnology is ephemeral, but within the gesture of touch is inscribed a potential deeper understanding of our imprint in the world. Each work reflects a desire to incorporate the viewer into a discourse about the evolution of change that will be generated by advances in nanotechnology.

## Material agency

What I want to highlight in this chapter, through setting up an analogy with drawing, is the link between the graphite in the pencil when it is applied to the paper and the agency and signification of the carbon molecules in quantum computing and thinking. The humble materiality of the pencil used to visualize ideas now encapsulates a key synergy with some of the major breakthroughs taking place in communication technologies, computing and nano-fabrication. Each drawn line creates marks that contain the data generated by the artist: data that go beyond the real. The atomic point of energy which 'move[s] the point into a line, the line into a plane, and the plane into a spatial dimension' (Klee 1961) is crucial to our understanding of the significance of the mark. The artist's perceptions, thinking and ideas are distilled and visualized through the molecular process of the graphite being transferred to the paper. The data, which the viewer perceives as a drawing, possess all the attributes and multitudes of meanings that are imbued within the invisible substrate underpinning that drawing's uniqueness.

The recontextualizing of the agency of an artist's medium allows for a conceptual shift and articulates a social reflection on art practice in the area of nanotechnology. The AFM's cantilever resembles the tip of the pencil as it moves across the sample substrate's surface drawing, transmitting signals back to the photodiode. The process of thinking through drawing could be seen to mirror that of quantum computing where data is imbued inside the carbon molecules. The analogy links the materiality of the drawing medium clearly with that of the immaterial development, at a nano level, of a contemporary computer. The same carbon used in both processes creates an indexical link that can reconfigure our perception of contemporary mark making from a completely different viewpoint. The shift is in our conscious understanding of a material's agency, challenging boundaries, territories and dimensions of being human. Nanotechnology research forces us to reflect and reconfigure our perception of self as matter in relation to all other material and immaterial things, including thought.

## Notes

1 In his book *Art and Agency* Alfred Gell wrote that agency is indexical to materiality that generates a series of meanings. Agency can work as mimetic, analogous and metaphorical in describing materials' possible meanings.

2 Nanoart is one of the new art disciplines developing at the intersections between art-science-technology.

3 The term 'nanotechnology' was coined by Norio Taniguchi in 1974. K. Eric Drexler is credited with introducing the term into public realm through his influential book *Engines of Creation*, 'We can use the terms "nanotechnology" and "molecular technology" interchangeably to describe the new style of technology. The engineers of the new technology will build both nanocircuits and nanomachines', 'Engines of creation the coming era of nanotechnology', http://e-drexler.com/d/06/00/EOC/EOC_Table_of_Contents.html. Retrieved 11 November 2007.

4 In this text I use 'immateriality' to define the new material condition that media artists are working with. The word is used in the context of this chapter to create a link with Nanoart through the de-territorialization and re-territorialization of the art object. Christiane Paul states that the basis for immateriality is 'software, systems and networks'. Paul, C. (2007), 'The myth of immateriality: presenting and preserving new media', *MediaArtHistories*, O. Grau. Massachusetts: MIT Press.

5 Democritus (460–370 BCE) was a Greek philosopher who was the first to name the atom and believed that 'genuine knowledge' could only be gained by thought and that the knowledge gained through the senses was 'obscure knowledge'. The mind and soul to exist cannot come from nothing and therefore they must be atoms.

6 Lucretius gave the name clinamen for the unpredictable swerve of atoms. According to Lucretius, the unpredictable swerve occurs 'at no fixed place or time'.

7 The Swerve of Lucretius comes from understanding the writing of Epicurus (341–271 BCE) in explaining the basis of matter. There is no real evidence of Epicurus's writings but his concept of the swerve is expanded by Lucretius's exploring his understanding that bodies are in motion and things do not come into existence from nothing. The bodies are compounds and as they move there must be a space (void) for bodies to move in. So the body as compounds of divisible parts that break down but not to nothing. The final parts are atoms that fall endlessly downward in the void where they unpredictably and randomly swerve. The swerve is a way of explaining how atoms join together to form macroscopic bodies and these bodies form larger bodies. Another reason for the swerve is to allow for individuality and human freedom to flourish denying control and dogma.

8 'The genesis of the individual and its defining characteristics one must assume the existence of a first term, a principle, which would provide a sufficient explanation of how the individual had come to be individual and account for its singularity

(haecceity), but this does not prove that the essential precondition of ontogenesis need be anything resembling a first term. Yet a term is itself already an individual, or at least something capable of being individualized, something that can be the cause of an absolutely specific existence (haecceity), something that can lead to a proliferation of many new haecceities. Anything that contributes to establishing relations already belongs to the same mode of existence as the individual, whether it be an atom which is an indivisible and eternal particle, or prime matter, or a form'. Simondon, G. (1992), 'The genesis of the individual', in J. Crary and S. Kwinter (eds), *Incorporations*, New York: Zone Books, 297–319.

9 A visualization of 'digital rain' was used in the opening credits to visualize the substrate of the virtual world created for the movie series *The Matrix*. The virtual world inhabited by the characters was constructed from computer code that only became visible due to glitches in the system.

10 The word 'pencil' comes from the Latin word 'penicillus', which means 'little tail'. This was the name of the tiny brush that ancient Romans used as a writing instrument. In 1795, Nicholas Jacques Conte patented a method of mixing clay with graphite to create different compounds and therefore different grades.

11 The van der Waals force is a very small electric charge that can hold materials together (like static electricity).

12 The discovery of the $C_{60}$ compound in 1985 earned Harold Kroto, James R. Heath and Richard Smalley the 1996 Nobel Prize in Chemistry.

13 'That pessimistic assumption was put to rest in 2004. One of us (Geim), in collaboration with then postdoctoral associate Kostya S. Novoselov and his co-workers at the University of Manchester in England, was studying a variety of approaches to making even thinner samples of graphite. At that time, most laboratories began such attempts with soot, but Geim and his colleagues serendipitously started with bits of debris left over after splitting graphite by brute force. They simply stuck a flake of graphite debris onto plastic adhesive tape, folded the sticky side of the tape over the flake and then pulled the tape apart, cleaving the flake in two. As the experimenters repeated the process, the resulting fragments grew thinner. Once the investigators had many thin fragments, they meticulously examined the pieces – and were astonished to find that some were only one atom thick. Even more unexpectedly, the newly identified bits of graphene turned out to have high crystal quality and to be chemically stable even at room temperature'. Geim, A. K. and Kim, P. (2008), 'Carbon wonderland', *Scientific American*, 298: 4, pp. 90–97.

14 The title can also be seen as a reference to the Gilded Age, a time of economic expansion roughly coinciding with the first Victorian era. 'The Diamond Age', *Wikipedia: The Free Encyclopedia*. Wikimedia Foundation, Inc. http://en.wikipedia.org/wiki/The_Diamond_Age. Retrieved 12 July 2010.

15 James Gimzewski used a Scanning Tunnelling Microscope to push atoms around. In 1996, as a scientist at the IBM Research Division's Zurich laboratory, Gimzewski built an abacus with individual molecules as beads with a diameter of less than one nanometre (one billionth of a metre).

16 Cartesian logic comes from René Descartes (1596–1650), a French philosopher who believed that people, the world and the universe all work like the mechanism of a clock as though everything was a well-oiled machine. For things to exist in space, space also had to be mapped and gridded. The space is where the atoms fall continually downward and there is no room for the swerve.

17 In his article, Colin Milburn references a series of reports written for the National Science Foundation, and the Department of Commerce funded workshop on human performance enhancement through convergence of nano with bio, info and cognitive science. Here a number of nano and bio scenarios are explored. Milburn referenced Goldblatt, M. (2002), 'DARPA's programs in enhancing human performance', *Converging Technologies for Improving Human Performance*, W. S. B. Mihail C. Roco, Dortrecht: the Netherlands, and Klumer. and Asher, R. (2002), 'Brain machine interface', *Converging Technologies for Improving Human Performance*, W. S. B. Mihail C. Roco, Dordrecht: the Netherlands.

18 James D. Watson and Francis Crick in 1953 worked out the famous double helix structure of DNA and were award the Nobel Prize in Physiology and Medicine in 1962.

# Chapter 2

## From Seeing to Touching: From the Invisible to the Visible

> The new tactile techniques opened up a radically new approach to microscopy enabling real local properties to be imaged and mapped. For instance, ultra high-resolution images of local magnetism like bits of north and south directed domains could be obtained with magnetic tips. If friction was an issue, images of local friction as it scanned the surface could be mapped. This opened up a new world, a world never really seen before on those terms – the nanoworld. Even bigger consequences of 'touching' rather than looking were also realized.
>
> (Gimzewski and Vesna 2003: 7)

This chapter explores the technical processes of the Atomic Force Microscope (AFM) and its relationship with the reinstatement of the significance of a sense of touch. The science of making the invisible visible through new technologies has been a constant source of inspiration for artists. The significance of drawing referenced in the previous chapter as being analogous to the AFM is revealed through its haptic interface with recording and revealing the immaterial and material world. The AFM creates images from the invisible by using a relatively simple haptic interface with the material relational fabric of the world, changing the meaning of 'seeing is believing'.

I want to look at a history of art and science, focussing on Etienne-Jules Marey (1830–1904), who by his research using the camera exposed the invisible, through to the imaging technology of the AFM developed in 1986. My aim is to establish a direct comparison between the Chronophotographic work of Etienne-Jules Marey at the turn of the twentieth century and the AFM's imaging of the atomic world.

Marey was a scientist working initially in the area of medical devices and later in the realm of image science. He revealed his unique role by using photography to scientifically demonstrate unseen aspects of the material world. The work that Marey and another well-noted photographer Eadweard Muybridge (1830–1904) produced in making what appears invisible to the eye visible had a profound effect on many early twentieth century artists. Marey's and Muybridge's work affected both the practice and the way in which artists responded to imaging the world and was particularly influential on the work of the Italian futurist and specifically Umberto Boccioni (1882–1916).

This chapter positions Boccioni as a touchstone who created in his sculptural works, visual and physical manifestations of boundaryless states. Boccioni's sculptures intuitively reveal the first 3D physical forms that envisioned an art of nanotechnology. This historical exploration of Boccioni's work is contrasted with contemporary nanotechnological visualization tools. I will show how the AFM constructs a machinic understanding of a material's atomic particles and works as a contradiction to traditional microscopes by privileging haptic oversight, thereby reintroducing a sensorial-machinic means of perceiving the world.

## The Atomic Force Microscope

The AFM, invented in 1986, was developed at the end of a long line of optical lens-based microscopes first introduced in the early 1600s. The history of imaging devices, to record the infinitely small invisible world, starts with the microscope. Its evolution was a synchronistic process that involved a number of researchers including Anton Van Leeuwenhoek of Holland (1632–1723) who is credited with being the father of microscopy. Robert Hooke (1635–1703), in England, enhanced the optical capabilities of the microscope and wrote *Micrographia*, the first book describing and illustrating observations made through the microscope.

In *Micrographia,* Hooke demonstrates the power of observation and analogy when viewing a sliver of cork. He notes:

> [the] pores, or cells, were not very deep, but consisted of a great many little Boxes (sic), separated out of one continued long pore, by certain Diaphragms, as is visible by the Figure B, which represents a sight of those pores split the long-ways.
> (Hooke 1665)

The drawn illustration (Figure 1) by Hooke shows the use of graphite in a highly experimental form of representing the thinking and interpretations of his perception of cork through a microscope. Hooke uses analogous language to try to translate what he sees and recognizes of the material world, through the microscope. He is said to have coined the word cell by relating what he saw through the microscope to a monk's cell, thereby demonstrating the power of a narrative when observing what has been perceived as invisible.

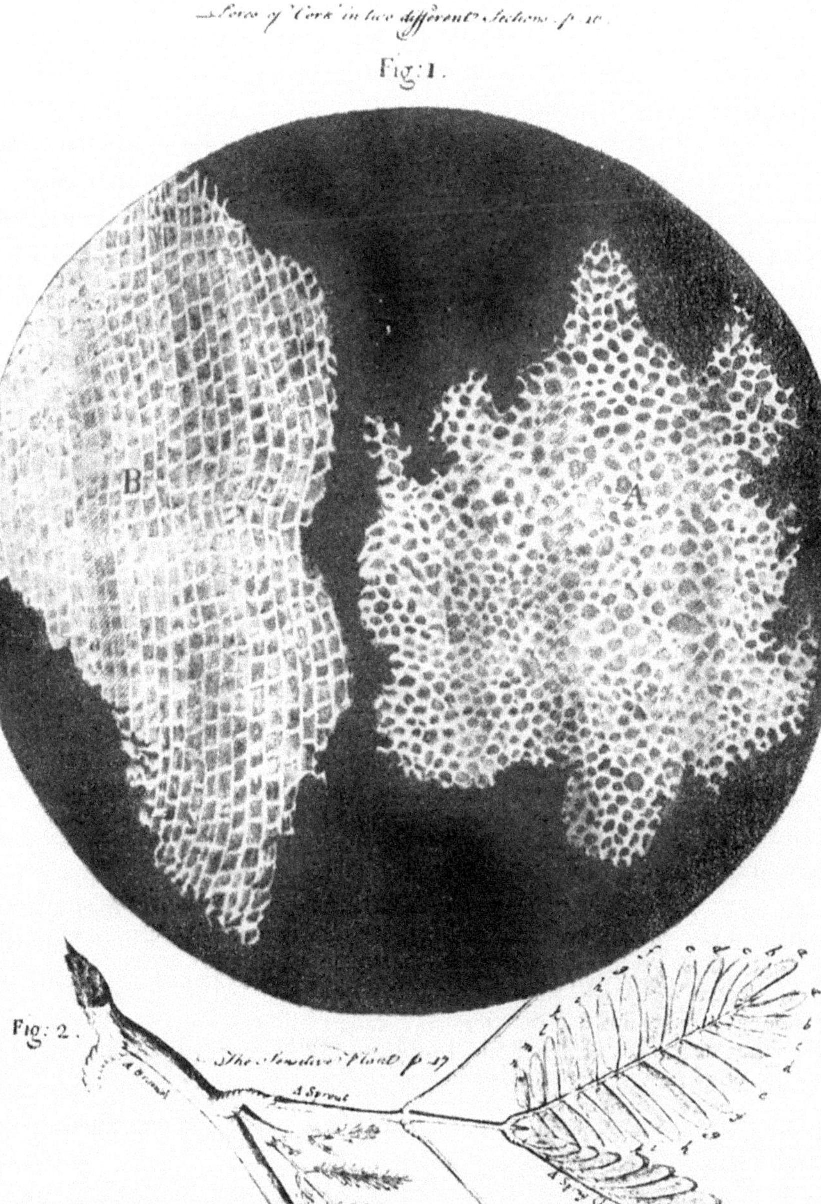


**Figure 1:** Robert Hooke, Suber cells and mimosa leaves, *Micrographia*. Figure A and B of the XI. Scheme for the Interstitia (1665).

Optical microscopes were used exclusively until 1931 when the first electron microscope was introduced. A number of significant discoveries were made that enabled the development of the electron microscopes and ultimately the AFM.

The electromagnetic spectrum discovered by James Clerk Maxwell[2] (1831–1879) gave a breakdown of the visual spectrum and demonstrated the range under which optical microscopes worked. Within this visual spectrum the range of visible light is between 390 and 750 nanometres. Red is at the low range of frequency and violet exists at the high-end of the frequency range.

Other important discoveries leading up to the development of the AFM included Wilhelm Conrad Röntgen's discovery of x-ray in 1895, Joseph Thomson's 1898 discovery of the electron, and Heinrich Hertz's clarification of electromagnetic theory of light as discovered by Maxwell. In 1900 Max Planck declared that energy is radiated in small, discrete units, which he called quanta, and in 1903 Richard Zsigmondy invented the ultramicroscope that could view particles below the optical range.

These discoveries were not only important for science, but they also formed a basis for artistic research at the turn of century in relation to cognition beyond human perception. As Linda Henderson explains: '[the] existence of invisible realms just beyond the reach of the human eye was no longer a matter of mystical or philosophical speculation; it had been established empirically by science' (Henderson 1983).

This shift from speculation to cognition, enabled by the use of scientific equipment, continued with the exploration of the infinitely small. Technological developments meant that previously unseen worlds could now be made a visible part of our conscious understanding. However, there were some obstacles to be overcome.

Optical microscopes are diffraction limited, that is, they can only image dark or strongly refracting objects effectively. The diffraction limits resolution to approximately 0.2 micrometre. Ambient light can diffuse the focus more than 1000 times better than the optical diffraction limit. To overcome the problems of working with light waves, a new method of quantifying the invisible was developed with the Scanning Probe Microscope (SPM) in the 1980s.

Unlike the previous microscopes, the Scanning Tunnelling Microscope (STM) family (which includes both the STM and the AFM) is not a lens-based vision system. These instruments rely on touch to create images that extend and redefine the concept of seeing.

The AFM uses a cantilever to see the world of atoms through a process similar to that of a stylus in an old-fashioned record player. However, the STM (the precursor to the AFM) recreates the topography by *almost* touching the sample's surface. The STM sharpened metal wire cantilever hovers above the surface of the substrate to be imaged by maintaining a distance separation of approximately 0.5 to

1.0 nanometre. This distance is maintained by a piezo element via the tunnelling current, which creates a precise distance between the apical atom on the probe's tip and the nearest atom on the substrate being scanned. The probe sends back information about its X, Y and Z movements across the surface, to software written specifically to interpret the data regarding the proximity of atoms in relation to one another and translates the information into what appears to be a topographical landscape.

The AFM differs from the STM in that the cantilever's tip, in contact mode, touches the substrates surface. There are three imaging modes that rely on a form of touch: contact, non-contact and dynamic mode or intermittent contact known by the industry name tapping. All these interact with the surface in different ways. The AFM and STM use a scanning process to obtain an image of the surface. The image is obtained by mechanically moving the probe, in a raster scan of the specimen line by line, and recording the probe-surface interaction via a laser beam reflected onto a photodiode. In contact mode the cantilever's tip makes direct contact with the substrate, whereas in non-contact mode it 'feels' the attractive and repulsive forces from the surface atoms without directly touching them. In tapping mode the oscillation of the cantilever is detected throughout the entire oscillation cycle of the sample surface.

With the AFM in non-contact mode, the system vibrates the stiff cantilever close to its resonant frequency (typically from 100 to 400 kHz) at an amplitude of a few tens to hundreds of angstroms. Changes in the resonant frequency or vibration amplitude are then detected as the tip comes near the sample surface. The sensitivity of changes in the surface recorded through touch or changes in resonant frequency provide sub-angstrom vertical resolution in the image (Benatar 1997).

When operating in force spectroscopy mode, the cantilever tip is brought down to the sample surface on the 'Z' coordinate. As the cantilever nears the surface of the sample, the van der Waals force steps in and jumps to contact, enabling it to record the surface atomic vibration and stiffness. In this mode the cantilever can be driven by a haptic interface that enables the user to sense the physical interaction and resistance of touching and releasing the surface. When the cantilever is removed from the sampled substrate, the haptic interface enables the user to feel and measure the force of attraction to the surface. While the imaging in these modes varies, each of the methods of data translation to image is relative to the intention of the research being undertaken.

Through these instruments of nanotechnology a fundamental change in our conscious understanding of materiality has evolved through the adoption of a quantifiable 'machinic' understanding of the material world. We now see the evidence of atomic imagery made visible by the instrumental tools of nanotechnology: the STM 1981 and later the AFM 1986.

The concept of gathering scientific data through touch via a cantilever allows for a reconfiguration of a dominant ocularcentric understanding of the world. The AFM is analogous to a prosthetic hand with its finger outstretched reading every tactile and sensory connection with the material substrate being examined. Once the finger has felt the surface the concept of nano assembly is assimilated; the STM and AFM are not only imaging devices, they are probes that can provide the potential to move atoms around, creating new materials, forms and structures replicating the biblical construction of humans from dust particles. This ability to move atoms was exploited by Don Eigler. He moved 35 Xenon atoms around to create the letters 'IBM'. Gimzewski used the carbon molecule $C_{60}$ to create a nano abacus.

The topographical landscape of the sampled substrate surface is a metaphorical connection that positions the scientist as an explorer fitting of the romantic view of the nineteenth century pioneers. These pioneers create an ownership via their gaze of what they surveyed across the new territory of the material world laid out before them. With molecular nanotechnology a new kind of ownership and colonization can be felt in the use of touch, which can be witnessed in when the AFM was used to move atoms spelling out the abbreviation 'IBM'.[3]

Reflecting on the photographic imaging work of Marey in the context of this chapter is a historical counterpoint to the AFM. Marey with his photographs presents us with a de-territorialization of the body, confronting space and time, and predicting a future of fluidity and mobility. The AFM gives us images that express the instability and immateriality of matter through patternation.

## Nanoart history

In an art historical context, shifts of attitude and consciousness can be linked with various artistic movements and 'isms' of the early twentieth century. The discourse developed by Arthur Miller (2001) in his book *Einstein, Picasso: Space, Time and the Beauty That Causes Havoc* hypothesizes a contextual biographical relationship between Picasso and Einstein even though they never meet.

> As we would have expected of those who defined the avant-garde, Einstein and Picasso were intellectual opportunists. They drew on apparently disparate fields, while working on the same problem – the nature of simultaneity. Einstein solved

> it for temporal simultaneity with the special theory of relativity and Picasso for spatial simultaneity in Les Demoiselles d'Avignon, which was the springboard to cubism. Both concluded that how you look at something, that's the way it is. There is no one true perspective.
>
> (Miller 2004)

The intuitive synergies of ideas, relative to movement, space and time generated at the turn of twentieth century, were being culturally filtered and explored by artists of their time. Miller develops an argument that situates a shift from visual representation of concepts to abstract concepts synonymous to those being explored and developed in mathematics simultaneously by both Einstein and Picasso. Kemp (1990), Petrie (1974), Braun (1992) and Cresswell (2006) are scholars who have explored these areas in relation to aspects of the cultural assimilation of science. Henderson in her article *Cubism, Futurism, and Ether Physics in the Early Twentieth Century* details how like Miller, scientific and philosophical findings were permeating artistic thinking and understandings simultaneously (Henderson 2004).

## Marey, Boccioni and Bergson

In focusing on the invisible made visible, I want to look at the Italian Futurist Umberto Boccioni's work between 1911 and 1914. The social effects of emerging technologies and science were important to the burgeoning relationships developing in the Futurist movement. Boccioni challenged traditional paradigms through his practice where he promoted and critiqued scientific and technological advances.[4] He had initially been inspired by the scientific work of Etienne Jules Marey's fixed plate chronophotographs created in the late part of the nineteenth century. Marey's technological chronophotographs reduced time to a point and made the invisible visible, thereby creating a new source of imagery for artists (Boccioni, Carrà et al. 1973: 15). The chronophotographic technique rather than style was unlocking the secrets of a fluid world, an essence or the concept of an ether. Boccioni perceived that between the binary of the fluidity of art and the methodologies of science there needed to be another influence and in this case he turned to the philosophical writings of Henri Bergson.

Significantly, Boccioni, who was initially inspired by modern scientific techniques of representation, rejected some of Marey's photographic influences as being too mechanical in their division and segmentation of time. Instead, he wanted to incorporate Bergson's philosophy which he saw as being integral to his concerns, and as an alternative position for coming to terms with current research expressed through science. The works produced by Boccioni reveal a hidden dimensionality of the universal, presenting an intuitive understanding of the inclusivity of the 'mathematical and geometrical'. 5

However, there were differences between Boccioni's and Bergson's positions. Whereas the philosopher made distinctions between intuition, 'that grasp the spirit', and the intellect, 'that grasps immobility and matter', the artist made no distinction between spirit and matter (Orban 1997: 58). The vision of matter, being part of an 'indivisible whole' that exists through continual vibration taking place at the atomic core, is experimented with, reflected upon and represented intuitively in Boccioni's sculptural abstractions.

## Etienne Jules Marey

While Boccioni's approach embraced the intuitive, Marey drew on scientific methods. He was continually looking for ways, 'to extend the ability of observation in order that patterns of regularity, and eventually laws, could be established' (Cresswell 2006: 73 ). He believed that if something can be made visible, it can be measured and then it can be known and laws developed, and that invisible motion, once represented visually, could be codified in this way.

Marey was a physiologist, doctor, biomechanics engineer and inventor whose work was significant in developing a number of physical instruments used in the areas of medicine, aviation and cinematic and photographic exploration. Early medical instruments, such as the sphygmograph, can be seen to have a direct link to Marey's desire to take the invisible nature of movement within the body and make it visible. The example of the sphygmograph can be seen to parallel the operational workings of an atomic force microscope. The sphygmograph was a machine initially developed by Karl von Vierordt in 1854 to determine a patient's blood pressure, but was modified by Marey to give a visual printout of the pulse. The modification of the instrument was very simple. It comprised of a lever, with one end resting on the pulse point of the wrist and the other connected to a stylus, and a clockwork mechanism that moved a strip of smoke-blackened paper under the stylus at uniform speed, converting the pulsations into a fluid inscription' (Braun 1992: 310). We might draw parallel between the operational workings of the sphygmograph and an atomic force microscope, as discussed earlier in this chapter.

Marey's fixed plate chronophotographs were indicative of the desire to turn motion into quantifiable data for scientific development. In his book called *Movement*, Marey explains that '[the] original form of the chronophotographic apparatus was very simple. It consisted of an ordinary camera and lens. Within the body of the camera, in front of the plate, a fenestrated diaphragm was fixed' (Marey 1895: 55). In reducing time to points and making the invisible visible, the chronophotographs created a verisimilitude representation of the visual world in the name of science. Marey's chronophotographs were shot against a black background[6] suggested to Marey by the chemist Chevreul whose work influenced Seurat. The black backgrounds of Marey's chronophotographs present a vivid visualization of ether by default as a necessity for taking photographs of movement.

A number of chronophotographic devices and methods (geometrical and stereoscopic) of representation were shown through Marey's photographs in the 1900 Paris *Exposition Universelle* (World Exposition). It is here that the Futurists would have likely encountered and been excited by them (Apollonio 1973: 15). Evidence of this likely encounter is seen in the Futurist's Photodynamism of Anton Giulio Bragaglia, and the effects observed in various paintings of Carlo Carrà, Giacomo Balla, Gino Severini and Boccioni. In their movement paintings the Futurist artists rearticulate the way the camera captures the moving image on a single plate photograph. In addition, Marey's machinic visualization of the invisible also created different ways a machine might record life and opens up possibilities for Bergson to challenge life not as recorded data but as 'lived' experience.[7]

## Boccioni

The work of Boccioni is seminal in the representation of the invisibility of matter. It is based on a rejection of the scientific research of the time, aligned with the work of Marey, and is embedded with Bergson's philosophy on duration, movement and matter.

Boccioni was a critic of Marey's photographs and distanced himself from Photodynamism that offered a machinic understanding of movement. Boccioni's critique of the Futurist photographer Bragaglia's use of Photodynamism is referenced by Martha Braun who states that Boccioni was 'the most vehement antagonist of Bragaglia's' (Braun 1992: 310).

Boccioni had aligned his thinking with Bergson's philosophical theories of duration, where movement and intuition coalesce. Bergson had been dealing with concepts of duration that, when focussed on matter, recognized 'elementary vibrations' where matter almost vanishes, but not to nothing. Bergson suggested that the more that duration as vibrating matter became part of our consciousness, the more 'different parts of our being enter into each other'. These remarks relate to the very core of Boccioni's works of 1912.

Boccioni recognized Bergson's concept of 'lived experience', and his close study of Bergson's writings is reflected in his sculptural work post 1912. The crucial changes taking place in Boccioni's work is his elimination of cinematographic depictions of movement – that of capturing a single frame at a time. Boccioni utilized a more phenomenological view of space, which was based on Bergson's vitalistic principle of *duration,* in which time is not seen as a series of frozen *moments* but as a continuum. Boccioni's sculptures are three-dimensional, physical conglomerations of materiality and surrounding space, and are inclusive of sensorial experience. What is particularly evident is his representation of the disintegration and reintegration of the body.

Boccioni makes reference to the disintegration of the body. He cites Bergson in 'The Plastic Foundations of Futurist Sculpture and Painting manifesto of 1913', where he explains, 'Any division of matter into independent bodies with determined outlines is artificial' (Boccioni 1973b). Here Boccioni demonstrates his understanding of Bergson by relating to concepts of the artificiality of determined outlines in the construction of his visual form. In his article entitled 'Boccioni and Bergson', Brian Petrie suggests that for them 'the purpose of art is not the presentation, for whatever reason, of certain fixed configurations of matter, "images", to which a name and a function can be assigned, but rather to render transcendent truths about ontological status of matter' (Petrie 1974: 143). Boccioni demonstrated and developed evolving concepts that were 'merely implicit' in Bergson's writings, about the artificiality of the static division of space. Petrie suggests that Bergson and Boccioni with him extend this concept, he writes: 'the notion of "rest" is itself an abstraction. … and matter itself "envisaged as an indivisible whole, must be a flux rather than a thing"' (Petrie 1974: 144).

The practical work Boccioni produced in a short period of time (1911–1914) contributes to our ongoing understanding of ways in which we might be able to come to terms with what science and philosophy were telling us in the twentieth century. Boccioni demonstrated a keen interest in how new science, technologies and philosophies impacted on society. The role of other visualizing technologies such as x-ray was not lost on Boccioni who references the x-ray (1895) in relation to painting in 1910 in 'Futurist Painting: Technical Manifesto'. Boccioni asks whether we could

> still believe in the opacity of bodies, since our sharpened and multiplied sensitiveness has already penetrated the obscure manifestations of the medium? Why should we forget in our creations the doubled power of our sight, capable of giving results analogous to those of the x-rays?
>
> (Boccioni 1973d)

For Boccioni, the matter of objects and the elementary vibration are intuitively understood and allow for a total immersion with the world. Boccioni stated: 'Let us open up the figure like a window and enclose within it the environment in which it lives' (Boccioni 1973c). This definitive statement links with the concomitant thinking of ether in that period when matter was seen as an energy. Linda Dalrymple Henderson explains that

> with radioactivity and Le Bon's talk of matter dematerializing into the ether, the recently discovered Hertzian waves of wireless telegraphy (as well as x-rays) focused popular attention on the invisible, impalpable ether of space. Space was not thought of as empty in this period, and the terms space and ether of space are often synonymous in the written record'.
>
> (Henderson 2004)

In Boccioni's work we have the beginnings of a conscious visualization of matter in art as a flux, not a thing. In his art, Boccioni was trying to intuitively understand the scientific and philosophical nature of matter, which would not be made truly visible until the invention of the AFM. The link between Boccioni and Bergson both coming to terms with what is now current thinking around the permeability of matter, is significant here. This vision of matter being part of an 'indivisible whole' qualifies with the durational aspects of rhythm, tempo, vibration and pattern, and draws parallels with the continual vibration of matter at its atomic core.

Boccioni's ideas and intuitive abstractions tackle some of the most intriguing concepts about revealing all matter as part of an ineluctable whole. The works shift the conscious understanding of matter and breaks our ocularcentric understanding of the world replacing it with a sensorial and immersive one. The boundaries between objects are broken down, and all things at their elementary level interconnect and interact. Boccioni's drawings demonstrate this thinking through ideas of materiality, tracing, reworking and tracking the durational immateriality of matter. Boccioni states that 'an infinity of lines and currents emanate from our objects, making them live in the environment which has been created by their vibrations' (Boccioni 1973b).

The sculptural works of Boccioni reflect a spatial homogeneity in which objects are not independent of each other but are interconnected by what he called 'plastic dynamism' (Boccioni 1973a). Boccioni's intuitive sculptural works de-territorialized the object world and represent Bergson's theories of an interrelated durational spatiality that is now being made manifest by nanotechnological scientific imaging currently reflecting theories of vibration.

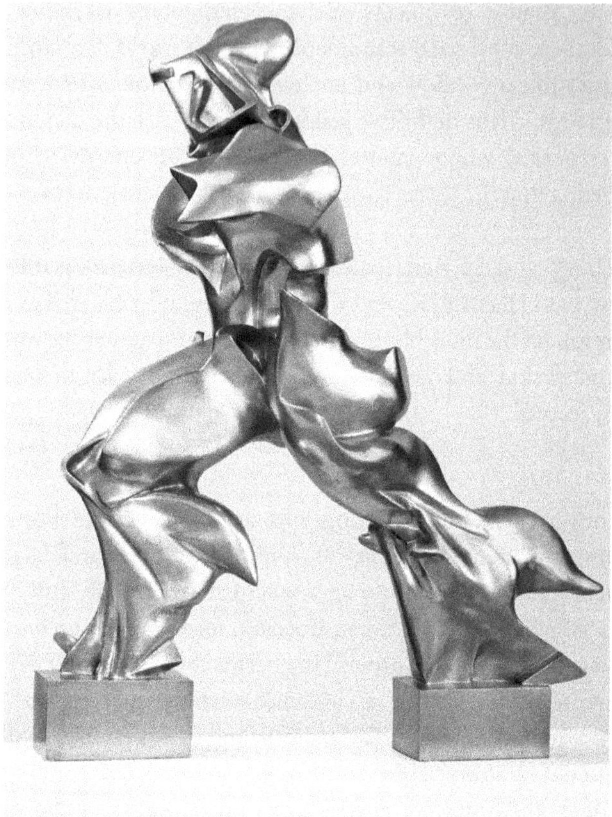

**Figure 2:** Umberto Boccioni, *Unique Forms of Continuity in Space* (1913). New York, Museum of Modern Art (MoMA). Bronze (cast 1931), 43 7/8 × 34 7/8 × 15 3/4' (111.2 × 88.5 × 40 cm). Acquired through the Lillie P. Bliss Bequest. 231.1948© 2012. Digital image, The Museum of Modern Art, New York/Scala, Florence.

Boccioni reconfigures our understanding of matter, not just a collapse of perspectival space. The body is not shown as 'isolated from but embedded within the surrounding space' (Thomas 2009b).

Boccioni's concept of objects absorbing their spatial surroundings can also be seen in relationship to his works that attempt to draw ideas back to a non-Euclidean way of conceiving space. *Unique Form of Continuity in Space* (Figure 2) confronts the gaze, exploring the sensing of space around and within the body (Thomas 2009b). If Boccioni was exploring these concepts in today's world of scientific research the ether of the early part of twentieth century might, in today's terms, be replaced by the concepts of Dark Energy and Dark Matter.

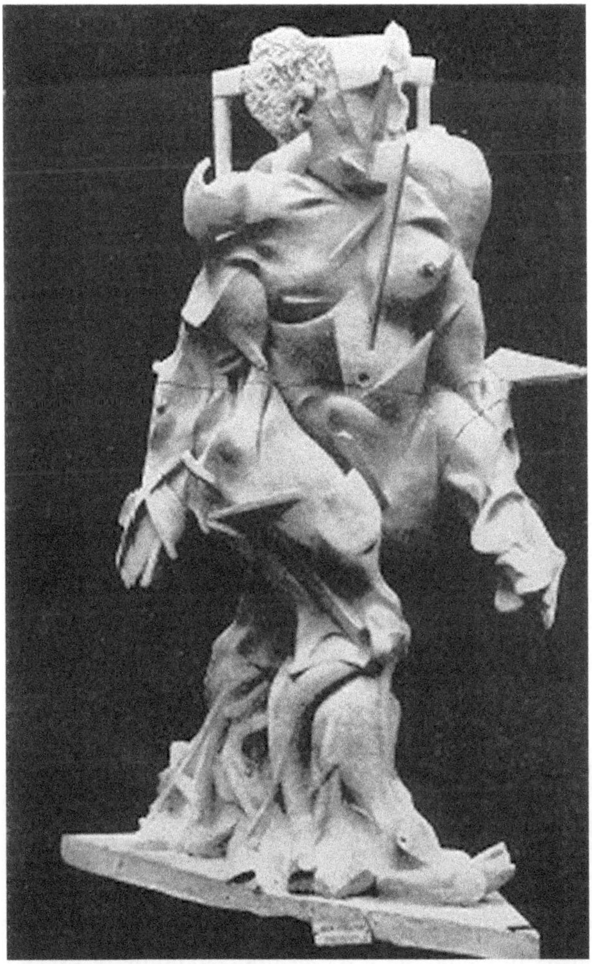

**Figure 3:** Umberto Boccioni, *Synthesis of the human dynamism* (1912). Destroyed sculpture.

In the *Synthesis of the human dynamism* we find that the depiction of ether is combined not only with an x-ray vision of the world but also a coalescence of all material objects. This is a remarkable synthesis of a world where organic and inorganic matter is as one. All elements in the work are non-hierarchical. The spatial alienation of perspectivalism was seriously contested by the work of Boccioni as space and time depicted through duration placed de-territorialization at the heart of his work. But Boccioni did so much more than contest the 4D space of cubism. He offers a new understanding of materiality. This 'extension of objects' that Boccioni addresses is seen in today's world

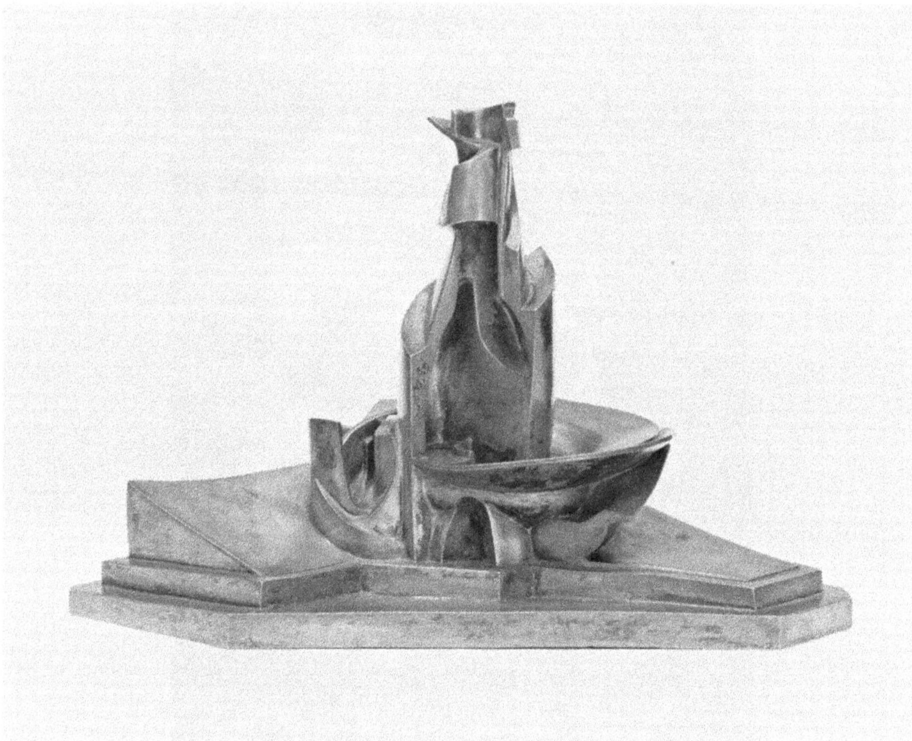

**Figure 4:** Umberto Boccioni, *Development of a Bottle in Space* (1912). New York, Museum of Modern Art (MoMA). Silvered bronze (cast 1931), 15 × 23 3/4 × 12 7/8' (38.1 × 60.3 × 32.7 cm). Aristide Maillol Fund. 230.1948© 2012. Digital image, The Museum of Modern Art, New York/Scala, Florence.

of nanotechnology as physical boundaries that become less discernable. Boccioni's work explored at its fundamental level an understanding of every atom's effect on every other atom and their influence on all material things.

An atom's effect on every other atom can be seen evolving in Boccioni's *Development of a Bottle in Space* (Figure 4), which illustrates a nanotechnological approach to matter and seemingly a 'bottom-up' creation of objects via the concept of nano assemblers. The evolution of the object is seen in Boccioni's work as a 'becoming' not independent and autonomous of the world but directly related to 'the environment in which it lives'. Boccioni's work stands as an example of his visionary comprehension and aims to create art that ruptures our understanding of matter and life. A direct comparison can be drawn between his work and Bergson's view of life, where 'the intellect, like the senses, is limited to taking, at intervals, views that are instantaneous and by that very fact immobile of the becoming of matter' (Bergson 1911: 288).

Boccioni demonstrates in the work *Development of a Bottle in Space*, the de-territorialization that confronts concepts of spatiality and duration being developed at the point in history prior to First World War. The subsequent events that took place in the First World War (1914–1918) saw the complete de-territorialization of the body witnessed first-hand in trench warfare. Similarly, the 1934 painting *Flanders* by Otto Dix presents boundaries of the fragmented bodies mingled with the mud, reflecting an outcome of a mechanistic causality when applied to humans. This causality is also reflective of what Eric Drexler called 'gray goo' which would be created by self-replicating nanobots that become autonomous and de-territorialize the materiality of the world (Drexler 1986).

## Nanotechnological scale

To define a contemporary understanding of matter, Colin Milburn positions the inception of nanotechnologies as a scientific discipline that: 'provokes the hyperreal collapse of humanistic discourse, puncturing the fragile membrane between real and simulation, science and science fiction, organism and machine, and heralding metamorphic futures and cyborganic discontinuities' (Milburn 2004: 123).

Thus the possibilities of metamorphic and cyborganic discourses being part human, part the surrounding space and part technology aligned some of Boccioni's spatial artistic concerns. For him the boundaries of the body become a signifier of an intuitive imagined particle relationship with the space between and enveloping us. This view has links with contemporary Nanoart practices. The making of the invisible visible via the AFM will be brought into the contemporary context with discussion of *Nano-Scape* by Christa Sommerer and Laurent Mignonneau and the *Midas* project by Thomas and Raxworthy.

## Nano-Scape

In the 2002 *Science + Fiction* exhibition at the Sprengel Museum Hannover and ZKM, Karlsruhe, Sommerer and Mignonneau presented *Nano-Scape*. The project was an attempt to bring nanotechnologies into cultural discourse and public awareness. The artists describe their *Nano-Scape* project as combining

> the research areas nanotechnologies, haptic user interaction and self-organising systems. Our goal was not so much to show pure data or facts but to let users intuitively experience aspects of nanotechnology through a haptic user interface and to show how intricate and complex interactions on a nano-scale level can be.
>
> (Mignonneau and Sommerer 2010)

This nano-scale sculpture is invisible, just like the nano-world it comments on. While science and media try to capture images of these tiniest of particles in order to understand their properties, *Nano-Scape* tries to make this nano-world intuitively accessible through touch (Mignonneau and Sommerer 2005).

*Nano-Scape* demonstrates, via its use of touch, the AFM's use of haptic forces in obtaining its representational effects. The work consists of a glass tabletop with a simulated array of 120 atoms on its surface that the user interfaces with by wearing a series of magnetic rings on their fingers. A camera tracks the rings as the hand moves over the surface. Positional data is sent to the computer where it is processed to create a simulation of the interactive forces, repulsion and attraction between the atoms (Mignonneau and Sommerer 2005).

Sommerer and Mignonneau used the modelled set of atoms to create a simplified metaphorical version of the van der Waals force. In this way the simulated atoms in

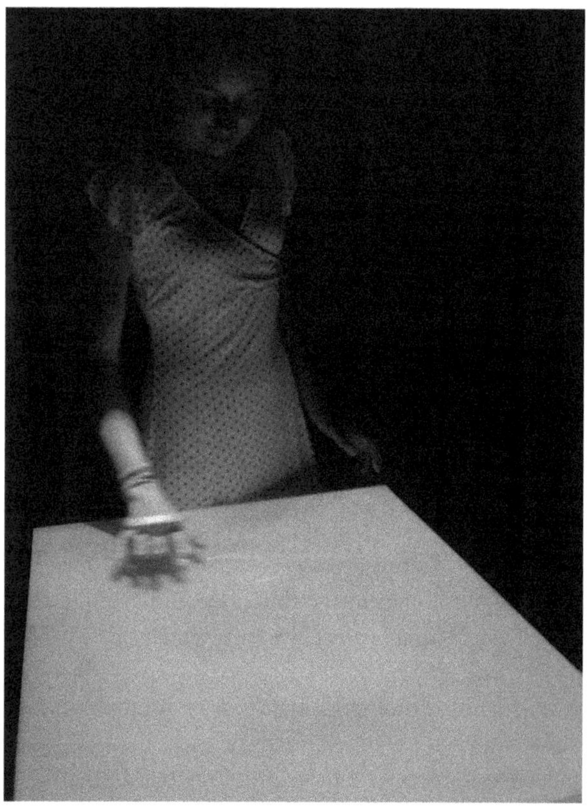

**Figure 5:** Christa Sommerer and Laurent Mignonneau, *Nano-Scape: user 06* (2002). Courtesy of the artists.

*Nano-Scape* are maintained in a state of equilibrium with their neighbours. The user, with their hand, simulates the cantilever used in the AFM to explore a haptic world of atoms. The user's hand interfaces at the very core of the vibratory forces (Figure 5) that draw and hold together to create the existence of matter. *Nano-Scape* seeks to define a context similar to that described in the work of Vesna and Gimzewski, where the indivisible world of atoms is turned into a physical interaction and users have some experience of atomism. We might suggest that both works highlight the new sensorial tactility that is needed to contest our ever-growing desensitization of our relationship with the material world.

## The Midas project

In my personal research practice that focused on the area of nanotechnology the initial project *Midas* (2007) explored concepts of touch using data recorded from the AFM at the Nanochemistry Research Institute at Curtin University of Technology. The concept of finding how much of us becomes part of the material world around us via touch was explored using skin cells, which were cultured during a SymbioticA residency in 2005 at University of Western Australia.

The *Midas* project is analogous to the curse of the fabled Midas, King of Phrygia, to whom Dionysus gave the power of turning all that he touched into gold. This gift changed the King's pride to a curse, as everything he touched, even his food and drink, were transformed into the precious metal. The king lost his ability to sustain himself emotionally and physically and asked the god Dionysus to remove the curse. This analogy provides a context to the inferences of nanotechnology and the alchemical transformation of atomic particles. The *Midas* project presents what happens to a skin cell on a nano-scale by creating a visual metaphor for revealing the de-territorialized and re-territorialized nanobiological body.

The installation consisted of a data projection and audio work created using the AFM in contact mode (as discussed earlier in this chapter). The contact mode was used to 'probe topology, conformational changes and interactions of a gold-coated cantilever on a single [skin] cell compared with an uncoated cantilever. Scanning the [skin] cell with both a gold-coated and an uncoated cantilever tip allowed us to record specific comparative data of each event' (Thomas 2009a).

The recorded sets of data from the AFM in force spectroscopy and contact mode were processed through software applications, translating the relevant information into sound files and image files.[8] Utilizing subsonic speakers, the data from the infinitely small atomic vibrations of the skin cell were made audible and palpable. The installation presented what happened at the point of transition when the first

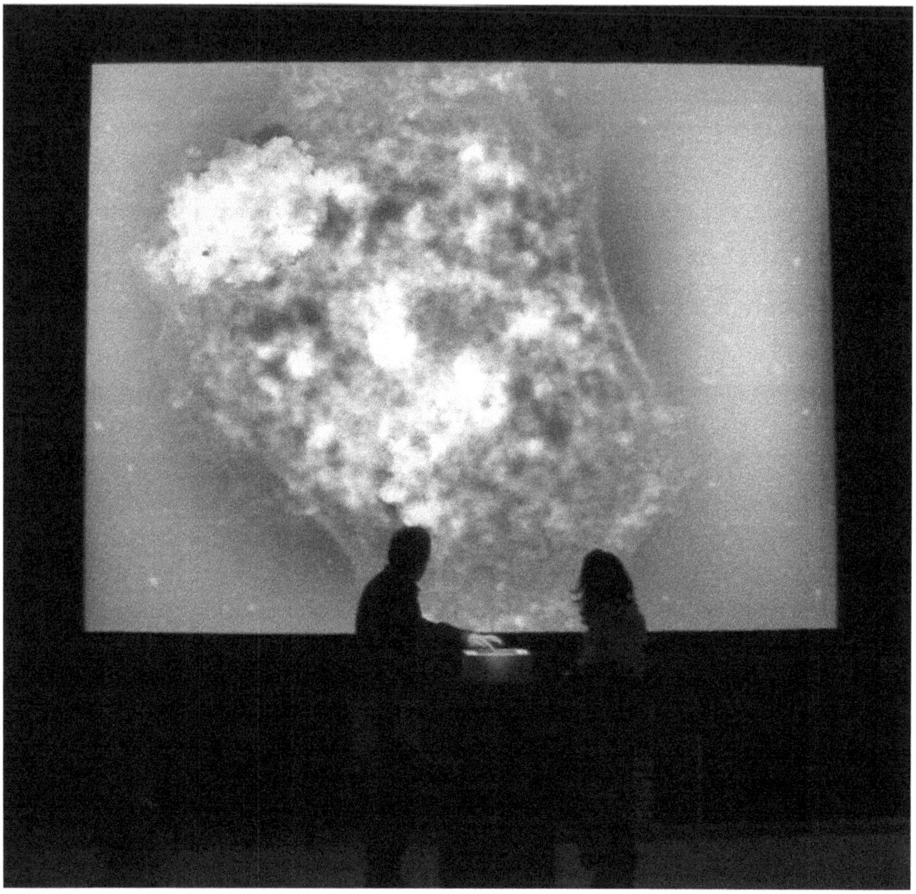

**Figure 6:** Paul Thomas and K. Raxworthy, *Midas* (2007). Still from video of the *Midas* installation at the Sk-interfaces exhibition closing event, FACT, Liverpool.

atoms of skin meet the first atoms of gold. The *Midas* project installation used the recorded data of vibrating atoms sonically, in conjunction with a genetic algorithmic visualization of the projected skin cell developed by Kevin Raxworthy. The genetic algorithm software was written to contaminate the image of the skin cell, replacing a Drexlerian de-territorializing landscape for semi-autonomous nano assemblers.

In *Midas* the viewer's touch (Figure 6) creates a sonic link with atomic vibration and is causal in allowing the release of semi-autonomous nanobots that eat and replace the suggested material base of the AFM's imaging of the skin cell, transmuting it into gold (Figure 7). The experience of touch is represented in this

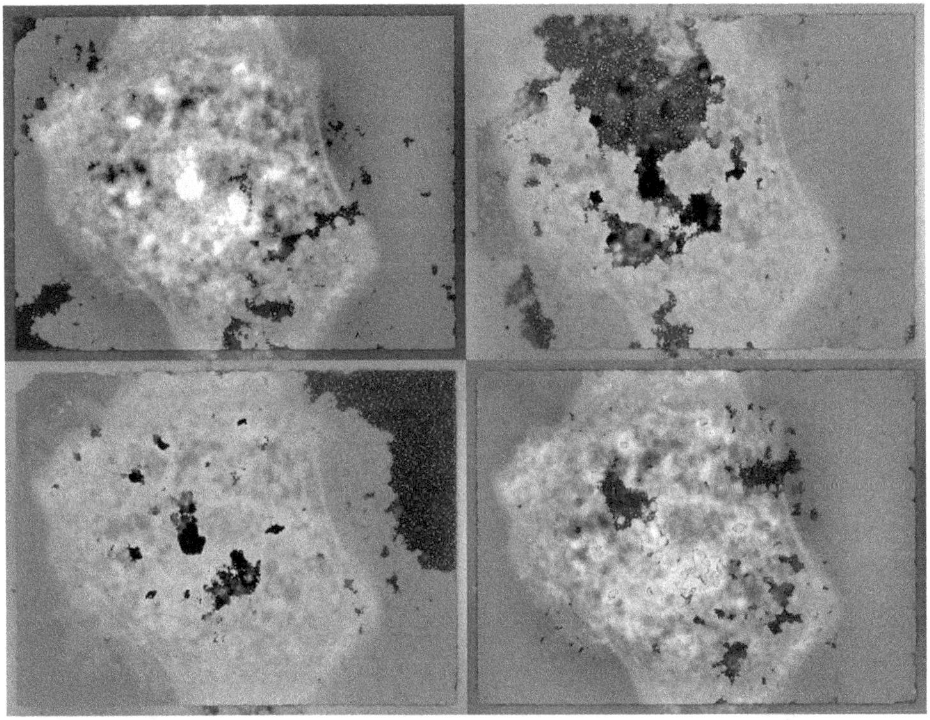

**Figure 7:** Paul Thomas and K. Raxworthy, *Midas* (2007). Four captured screensaves from the *Midas* installation.

process via the viewer making contact with a 9-carat gold-coated metal skin cell constructed from a 3D plotted image. The digital sound for the installation presents the viewer with the AFM's tactile analysis as an audible topographic map where speakers amplify the data of the atoms' vibrations.

Colin Milburn in reviewing the work suggests:

> Midas project renders palpable the disintegration of corporeal boundaries at the nanoscale, the electronic and atomic exchanges that constitute the zone of touch, the particulate tags or tactile transactions that happen in the interface. By incorporating the viewer into this molecular commons – the atomized zone of metal and flesh held in common – Midas creates a temporary, even instantaneous, space of intimacy and embodied practice (a nano-practice of contact), which facilitates critical reflection on the transvaluation of both the human and the atom in the global discourse of nanotechnology.
>
> (Milburn 2010: 18)

The intimacy and sensation of where the boundary of the body finishes and the world begins is highlighted in the project *Midas*.

## Bottom-up technology

In the nano-world, materiality can be de-territorialized and re-territorialized. The re-territorialization of materials uses atoms that are reconstructed using bottom-up technologies. Nano assembly, or the 'bottom-up' approach, is the construction of a supramolecular chemistry by the assembly of nano-scopic particles, or even atoms and molecules. If matter and nature can be constructed by bottom-up technology, then what constitutes how things look, their legal status, their agency and their role within society? All aspects are up for questioning. To illustrate this concept, the reconstruction of materials can be seen metaphorically in the same way that binary code is based on a machinic de-territorialization and re-territorialization of data by digital devices. A digital device such as a camera interprets and processes the material world into a digital manifestation of a thing. The image is captured into a series of pixels that are predetermined and locatable through code. The pixel pre-exists the image taken as the software in the camera seeks to assimilate the world. When each appropriate colour pixel is recoded it forms the illusion of image captured. The de-territorialization and re-territorialization of matter is the basis for a new conscious understanding of our human relationship to the material world (Deleuze and Guattari 1987: 347).

In the potential reduction of art to either the machinic binary code or to an atomic molecular level, we re-question the materiality of the artefact and where imbued meanings might reside. Art that explores and critiques nanotechnological research questions the substrate, tools and techniques used in the making of an artwork and the imbued meaning underpinning its signification.

## Nanotechnology and materiality

Roy Ascott is a pioneer of new modalities in the creation of art and scientific research, specifically in the role of understanding materiality and immateriality. He regards the relationship between scientific research and artistic practice as paramount in the evolution of crossovers between modalities of practice. He states:

> It is clear to me that research in biophysics, not least in the area of biophotonics, and that of electromagnetic fields, will play a significant part in the evolution of moist

> media, the substrate of 21st century art, embracing crossovers between telematics, neuroscience, molecular biology, quantum physics, and nano-engineering in the work of artists, designers, performers and architects.
>
> (Ascott 2004)

The translation and transmutation of the material world via nanotechnologies is emphasized through the immateriality and ephemerality of the data collected from the STM and AFM. The ephemerality of new nanotechnologies has demonstrated that there is a need for a different approach to the concepts and representation of materiality when linked only to scientific instrumentation. The imaging taken by AFM is based around data collected via a series of phases of the cantilevers touching the sample. The data is then used to create a proximity chart that when processed by the tinting of the embedded software colour chart creates a visualization of the code.

The reinstating of touch via the AFM as an aid to making the invisible visible comes at a time when we are seeing shifts in contemporary art practice towards a techno-mediated re-enchantment with the sensorial body. As Mark Hansen points out:

> New media artists directly engage the bodily dimensions of experience that surface, as it were, in response to the automation of vision. Their work can thus be said to invest the 'other side' of the automation of vision – the affective source of bodily experience that is so crucial to reconfiguring the human perception in our contemporary media ecology.
>
> (Hansen 2006: 101)

Hansen reveals the media artist as one who takes the automation of machine vision and confronts it as a modality of proprioception that he calls 'affectivity'.[9] The sensory nerve endings within the body provide the sense of the body's position therefore reconfiguring our perception of spatiality. The constant desire of reconfiguring human perception is now shifting away from Hansen's thinking towards a reconfiguration of materiality. Here we have new technological machines developing images that only simulate reality being 'real'. The scientific nanotechnological imaging equipment (AFM) creates new machine visions that construct 'real' images of the world's material and immaterial substrate.

The position and material substance of the body in space questions its physical boundaries in the same way that Boccioni's *Dynamism of a Speeding Horse + Houses 1912* presented the coalescence and coexistence of objects, sound, light and smell.

Nanotechnology prompts us to look at what constitutes the infinitely small, questioning the definition of nanotechnology art relevant to a measurement that is established through scale that starts at a measurement of 100 nanometres. The artist's conception of atomic is not restricted by the category of the atomic world beginning at 100 nanometres but as a stimulant for the redefining of material attitudes.

To define a contemporary understanding of matter, Colin Milburn states that the inception of nanotechnology as a scientific discipline 'provokes the hyperreal collapse of humanistic discourse, puncturing the fragile membrane between real and simulation, science and science fiction, organism and machine, and heralding metamorphic futures and cyborganic discontinuities' (Milburn 2004: 123).

The works of the artists referenced above critique current scientific research in the area of nanotechnology revealing a contested space of enquiry. The energy of vibration, pattern and rhythm is at the base the connectivity of matter that is translated by these artists in to visual experiences, electromagnetic sensations and sonic topographies. These sensory experiences reveal to the human body intimate understandings of the nano-world as a lived experience. The concept posed by Gimzewski of our being able to hear the well-being of the world by listening to atomic vibration through the prosthetic touch of the AFM, expands our awareness of all matter as a vital force. Tuning in to the world of atoms needs to be explored with lightness of potential for 'becoming'. Physical expression of force, like Serres' atoms falling regimentally to the ground destroying the lightness of touch, hearing, sight, taste and smell, can act as a barometer for the well-being of all matter organic and non-organic.

Nanotechnological research raises questions about what constitutes life itself below a cellular level. Thus the possibilities of metamorphic and cyborganic discourse being part-human explores the assimilation of the human and the material worlds. In this assimilation, the boundary of the body revealed through nanotechnology becomes a signifier of an imagined particle relationship with the space between and enveloping us. The human body as a material entity has its boundaries reshaped by the metamorphic assimilation of technologies. Therefore our understanding of the material world shifts to construct different agencies for our interpretation of matter.

## Notes

1 It was Robert Hooke (1665) who published 'Micrographia' where he coined the term 'cell'. When looking through the microscope at box like cells of cork he was reminded of a monk's cell in a monastery. Hooke, R. (1665), *Micrographia Some Physiological Descriptions of Minute Bodies Made by Magnifying Glasses with Observations and Inquiries Thereupon*, London, Royal Society.

2 James Clerk Maxwell's most prominent achievement was the putting together of three previously unconnected theories of electricity, magnetism and optics. The one constant in his equation was the speed of light suggesting that light is part of an electromagnetic field and travels through space in waves.

3 In 1990, the International Business Machines (IBM), a leading US computer manufacturer employees Don Eigler and Erhard Schweizer used the AFM to construct the company logo with pointy bright-blue xenon atoms on a smooth dark-gray nickel surface.

4 In this book I do not want to delve into the Futurists' political beliefs except to point out that the movement was aware of its time and incorporated social, technological and scientific understandings into its practice.

5 An extract from 'Technical manifesto of Futurist sculpture' by Boccioni states that a 'sculptural composition will contain in itself the marvelous mathematical and geometric elements of modern objects. These objects will not be placed alongside the statue, like so many explanatory attributes or separate decorative elements but, following the laws of a new conception of harmony'. Boccioni, U. (1973), 'Technical manifesto of Futurist sculpture', *Futurist Manifestos*, U. Apollonio, London: Thames & Hudson, pp. 51–65.

6 Marey states 'The nearest approach we have been able to make to these ideal conditions of Chevreul was by constructing a dark and capacious shed at the Physiological Station, the interior of which has been painted black, and by hanging a black velvet curtain at the back … Theoretically, an indefinite number of images may be taken in front of a dark background without any impressions of outside objects appearing on the plate'. Marey, E. J. (1895), *Movement*, New York: D. Appleton and Company.

7 Bergson suggests when discussing the lived explains that 'It coincides with my impatience, that is to say, with a certain portion of my own duration, which I cannot protract or contract as I like. It is no longer something thought, it is something lived'. Bergson, H. (1911), *Creative Evolution*, New York: H. Holt and Company.

8 When in 'force spectroscopy' mode, the AFM lowers the gold-coated silicon nitride tip to one location, touching and reading the surface vibration of the atomic particles. The cantilever in this context is like a metaphorical finger touching a surface to feel what is happening (like feeling one's pulse).

9 Proprioception refers to the unconscious perception of movement and spatial orientation arising from stimuli within the body itself. Proprioception is the sense of position and movement of the limbs and the sense of muscular tension.

# Chapter 3

## Nanotechnology, Vibration and Vitalism

The chapter explores ideas of atomic vibration and the life forces that are associated with the ongoing oscillations of atoms at the heart of all matter. Some claims made in the name of nanotechnology concerning what constitutes life and living are explored and contested. In this chapter, I explore examples of the nanotechnological rhetoric to examine where the language and science position the study of matter and how it is critiqued through artistic practice. In Chapter 2, I discussed the work of Boccioni, who with Bergson, created a view of matter that has a history of unpredictability linking back as far as Epicurus and Lucretius.

Bergson laid the basis for such a discourse on nanotechnology at the turn of the twentieth century in *Matter and Memory* (1896) and *Creative Evolution* (1911). In *Matter and Memory* Bergson stated that the dialectic between the machinic (measured movement in space) and the sensory are

> two different worlds, incapable of communicating otherwise than by a miracle – on the one hand, that of motion in space, on the other hand, that of consciousness with sensations .... Between quality on the one hand and pure quantity on the other.
>
> (Bergson 1990: 202)

The space between the machinic (quantitative) and the conscious (qualitative) in the atomic realm can be understood as either atoms falling in a regimented fashion or through the embrace of the 'clinamen' and the unpredictability of the swerve.

Thus the human becomes an example of the dichotomy between life and matter, spatially the body when viewed at an atomic level is both unpredictable and machinic. The body in this context has no real defining boundaries. The space of the human body and all matter is constructed from the morphology of the swerve and seminal to that is vibration, rhythm and pattern (as discussed in Chapter 1).

The desire to examine where the essence of what we call life can be found and quantified is to understand a form of vitalism. Jane Bennett's research into vitalism and mechanism focussed on the work of Immanuel Kant (Bildungstrieb), Hans Dreisch (entelechy) and Bergson (élan vital)[3] to pursue a history of the evolving reinterpretation in the comprehension of matter. For Bennett: 'vital materialism posits the causality of both inorganic and organic matter to be, to some extent, inscrutable to us and also that a mechanistic model is inadequate to both' (Bennett 2010: 67).

In the quest for the understanding of all matter to have a potential for life, the critical points are identified in Bergson's concept of élan vital where, he suggests, matter is infused with its own innate powers. In this concept he creates a differentiation between 'life and matter' that is based on duration and its connection to his interest in biology. Similarly, Bennett suggests that the vital materialist

> affirms a figure of matter as an active principal, and a universe of this lively materiality that is always in various states of congealment and diffusion, materialities that are active and creative without needing to be experienced or conceived as partaking in divinity or purposiveness.
>
> (Bennett 2010: 93)

These evolving concepts called critical vitalism, that were being discussed at the turn of the twentieth century, can be seen in contrast with ideas that nanotechnological research is developing today. Nanotechnological rhetoric positions itself firmly in a world of the quantifiable, mechanical and regimented fall of the atoms. But, if we were to affirm that matter is vital in a fluid relationship with what is called living, and this relationship predates an anthropocentric view of the world, where does life reside below the cellular level and how can the nanotechnological rhetoric be reconciled with this.

In contemporary discourse matter becomes a point of focus and the idea of matter always being active in principal remerges through Bergson's concepts of élan vital and duration. Bennett in her discussion of matter builds on Bergson's 'elementary vibrations', where matter almost vanishes. For Bergson the more 'duration', seen as vibrating matter, became part of our consciousness, the more 'different parts of our being enter into each other' (Bergson 1911: 212). These concepts are at the very core of Boccioni's Futurist works of 1912 (as discussed in Chapter 2), where representations of the vitality of matter and its 'elementary vibration' were intuitively understood. The intensity of interest in a world that showed unforeseeable change and experimentation enabled Boccioni to reflect through his work a total mingling and connectivity. 'Matter thus resolves itself into numberless vibrations, all linked together in uninterrupted continuity, all bound up with each other, and travelling in every direction like shivers through an immense body' (Bergson 1996: 208).

The identification of vibration as core is in itself seen as a potential by artists as a force that causes insights into this 'immense body' that is the world. The materiality of the world and that materiality's relationship to life is seen in Deleuze and Guattari's reflections on metal and metallurgy, they suggest:

> What metal and metallurgy bring to light is a life proper to matter, a vital state of matter as such, a material vitalism that doubtless exists everywhere but is ordinarily hidden or covered, rendered unrecognizable, dissociated by the hylomorphic model. Metallurgy is the consciousness or thought of the matter-flow, and metal the correlate of this consciousness.
>
> (Deleuze and Guattari 1987: 411)

The materiality of metal is part of the industrial modernist revolution where metal as a commodity is significant in its purchase on our conscious understanding of stability within the world. The idea that metal has a latent potential property that is not revealed but 'doubtless exists' as a conscious relationship with and within matter seems to be at the crux of Deleuze and Guattari's comments.

## Fiction to fact

Having touched on vitalism, matter and introduced vibration I would like to return to the concept of the swerve by exploring its implications and its natural relationship with concepts of the swarm.[4] What Lucretius highlighted with his concept of the swerve, is that the change in an atom's angle of inclination independent of the gravitational force of decline, is relative to concepts of the swarm. The swarm demonstrates behaviours of self-organization in simple systems where reactions lead to the development of intelligence. Michael Crichton in his novel *Prey* fictionalizes the concept of a swarm in the context of the body, and suggests that we could think about the body in terms of a giant swarm. The story's narrator describes the body saying:

> more precisely, it's a swarm of swarms, because each organ – blood, liver, kidneys – is a separate swarm. What we refer to, as "body" is really a combination of all these organ swarms. We think our bodies are solid, but that's only because we can't see what is going on at the cellular level. If you could enlarge the human body, blow it up to a vast size, you would see that it is literally nothing but a swirling mass of cells and atoms, clustered together into smaller swirls of cells and atoms.
>
> (Crichton 2002: 374)

Crichton's description defines a way of perceiving the body below the cellular level in terms of semi-autonomous swarms of cells and atoms held together in a fuzzy milieu. These milieus have self-organization and they have a fundamental rhythm that has 'become expressive' (Deleuze and Guattari 1987: 315). The swarm becomes identified by its expression as a milieu by some form of its own identification, as part of a process that is becoming individualized.

The concept of envisaging the body as a series of 'expressive' milieus is expanded by Deleuze and Guattari. They state that as our awareness of the immateriality of matter is extended through our conscious understanding, the world becomes de-territorialized and we become molecularized. We are, they suggest 'a molecular population, a people of oscillators as so many forces of interaction' (Deleuze and Guattari 1987: 345).

In this world of oscillation and vibration we can no longer talk about the body or being embodied, we are rather an 'array of bodies, many different kinds of them in a nested set of microbiomes' (Bennett 2010: 112–113). The nested body when seen in this boundaryless state of milieus does not create a demarcation between human, animals, vegetables or minerals. We see that the internal vibrancy of all matter demonstrates some force that can become 'lively, affective and signalling', the signalling being part of the connectivity created by the unpredictability of the swerve (Bennett 2010: 116). In this state our organs are not contained within the body, but are integrated into the material environment as an array of visual, auditory or electric signals, the same analogous signals that are running down wires or transmitted wirelessly.

## The refrain

Deleuze and Guattari's philosophical writings on the *refrain* parallel Crichton's post-biological state of the body, the 'signalling' and the concept of each organ being conceived as a swarm of atoms. The fictional vibrating mass of swarming atoms referred to in Crichton's *Prey*, which are replicating body organs, can be seen to exist in a state *born* out of chaos. But, while swarms are potentially born out of chaos, Eugene Thacker suggests, 'a swarm may exhibit a discernible global pattern' (Thacker 2004). However, Thacker suggests that while a discernible pattern arises 'this does not mean that a swarm prioritizes the group over the individual. Because of this, a swarm does not exist at a local or global level, but at a third level, where multiplicity and relation intersect' (Thacker 2004). It is this concept of a level where multiplicity and relation intersect that parallels Deleuze and Guattari's progression from interconnecting states that emerge out of chaos, by the refrain, that creates a milieu that can be defined as a territory.

The organs as milieus interconnect. Deleuze and Guattari suggest: 'This is in fact an act that affects milieus and rhythms that "territorializes" them' (Deleuze and Guattari 1987: 314). In the case of a refrain, they state: '[it] is rhythm and melody that have been territorialized' and they continue as a contextualizing expressive force, 'rhythm is the milieus' answer to chaos' (Deleuze and Guattari 1987: 313).

Here rhythm is not repetitive; it is the difference in vibrations that communicate between milieus: 'the difference that is rhythmic'. The vibrations create rhythms that can move between internal and external milieus as interconnected synergies (Deleuze and Guattari 1987: 317).

The refrain highlights the link between metaphysics and science, creating a platform for the vibrancy of atoms to be at the core of territorialization. The refrain is born out of three basic components: from a centre to the home, to interaction with the outside world. For example, initially a bird's refrain confronts chaos by marking out its territory, the sonic vibration centres the bird creating a home or nest. The second component explores the bird's refrain as creating a milieu with a non-physical boundary. The third component of the refrain marks out a line of communication allowing for a migration between other milieus beyond the non-physical boundary (Deleuze and Guattari 1987: 311). If we are to look at what is, as Elizabeth Grosz suggests

> transmitted, transformed, and relocated in this movement of forces from chaos to milieu to territory and from rhythm to the refrain to music, is nothing but vibration, resonance. Vibration is the pulse running through the universe from its chaotic interminability to its most intimate inscription on living bodies.
>
> (Grosz 2006)

This chaotic interminability is not simply inscribed on our bodies but is part of the very nature of the forces that defines us at present on a molecular level.

## Milieus

The link made between the organs as milieus is where the organs' acts are components that each contains a refrain within the territorialized body. Milieus that have been territorialized are seen in relationship to nano-vibrations, existing as a swarm of oscillating atoms that compose the immaterial substrate of the world. The milieus as organs express themselves through intermediary connecting rhythms with other milieus that have become, in part, territorialized. The humanness of this territory is revealed through the expression in the refrain that maps out the individual significance of swarm to swarm, cell to cell, tissue to tissue and organ to organ within the territorialized body. The milieu exists as atomic; each atom is a milieu that contains the very essence of life itself. The fall of the atom and the innate potential for the swerve creates the scenario for not only life to exist but for it to evolve.

## Post-biological body

In the current conscious concept of the human, the bodies' extremities are constructed through a desire to maintain its objectification. De-territorialization allows chaos to be a contributing factor in the development of the self-organizing evolutionary body and its relationship with matter and of the world.

The post-biological body is a space of milieus within territories that can become a de-territorialized zone through a change in refrain. The de-territorialization of matter is referenced to weakening the attracting bonds that hold the atomic structures together in its various unique states. When bonds of attraction between atoms are weakened through the manipulation via *bottom-up*[5] technology, what was an original swarm constituted as the body now has very porous boundaries. In the world of nanotechnology bodies that can be *de-territorialized* could therefore be *re-territorialized.* The re-territorialization atom by atom can be constructed with bottom-up technologies binding elements together for molecular manufacturing. In the same way that binary code of digital culture is based on machinic *de-territorialization and re-territorialization* of data by digital devices, the machinic controls interpret and process the decoding of data into the manifestation of a thing as a co-conspirator, re-coding and re-translating. The encoded and decoded machinic interpretation implies a territory of continual re-translation as a reaction to chaos. Relating data to atoms opens up a cyborg relationship with the de-territorialized body, which can cause the body to be seen as a material property to be exploited. Thus nanoconsciousness focusses on what it is to be human, as changes to our fundamental understanding of our materiality is challenged.

The narrative of the post-biological body for Milburn is something that takes place, he suggests within the 'scene of disintegration'. The post-biological body, Milburn explains 'deterritorializes the components of the body and simultaneously destines the molecular machines of the living cell towards a future where "life itself" has been "reshaped"' (Milburn 2005: 284). Similarly, the scientist with a research agenda can implement slight shifts in the construction and reshaping of the atomic world that in turn affects the molecular construction reconstituting possible life. Thus, in Milburn we see the rhetoric of nanotechnology critiqued and positioned in relation to the post-biological body.

Milburn looks at a number of ways that nanotechnology has been culturalized from science fiction to fact and it is useful to examine these. For example, in Will McCarthy's science fiction novel *Bloom* a utopian molecular society is constructed from the de-territorialized humanity on earth (McCarthy 1998). This de-territorialized humanity generates a totality based on swarm intelligence. McCarthy's *Bloom* suggests that life is inherent within the molecule, existing below the cellular level.

Jesper Hoffmeyer, a biosemiotician, makes the point that 'the swarm in which intelligence manifests itself is exactly that entity we call the body' (Hoffmeyer 1994). The points that Hoffmeyer makes in his writings on swarm intelligence are that

> [b]iologically speaking, the body can be understood as a swarm of cells and tissues which, unlike the swarms of bees or ants, stick relatively firmly together. However, the swarm of cells constituting a human body is a very different kind of swarm from that of the social insects. The body swarm is not built on ten thousand nearly identical units such as a bee society. Rather it should be seen as a swarm of swarms, i.e., a huge swarm of more or less overlapping swarms of very different kinds. And the minor swarms again are swarm-entities, so that we get a hierarchy of swarms. At all levels these swarms are engaged in distributed problem solving based on an infinitely complicated web of semitic interaction patterns which in the end can only be explained through reference to the actual history of the body system, evolution.
>
> (Hoffmeyer 1994)

The distributed nature of consciousness via the swarm suggests that inherent in the collective particles are forms of life. The simple collection of cells belies what happens below a cellular level, where cells themselves are swarms of swarms. So we might ask what happens when we are confronted with the physicality and unpredictability of all vibrating matter and the absence of life.

Roy Ascott, who is a visionary in critiquing contemporary scientific research, draws our attention to nanotechnological rhetoric. Ascott envisions that the problem is not in relation to molecular robotics theoretical benefits to mankind but whether this step takes us closer to being 'spiritually hollow', he states:

> [M]olecular robotics, positional assembly, and self replication suggest exciting possibilities for moving atoms around, building new materials, manufacturing nano machines, and generally building the fundamental blocks of nature into any configuration we desire, there is a danger that the outcomes, even when beneficial in engineering, medical and social terms, could be spiritually hollow, and as such would exacerbate rather than relieve the excessive materialism of our time.
>
> (Ascott 2004)

At a time when advances in nanotechnology are evolving rapidly the fundamental changes that are taking place are via shifts at a subconscious level of the human-centric comprehension of the material world. The power to reflect, deduce and develop a critical stance towards a nanotechnological future is fading fast as we become a reflection of nanotechnology's constructed fiction. The final frontier, the next 'industrial revolution', it has been suggested is nano, as it becomes not the ultimate boundary in a quest for knowledge, but a replication of its scale and investment. However, for now our material world is confronted by nanotechnological science fiction that is becoming factual, believing its own rhetoric.

## Nanotechnological rhetoric

In 1999, a nanotechnology vision was formulated aimed at 'establishing a foundational knowledge at the nano-scale'. Economically this new knowledge 'underpinned about a quarter of a trillion dollars worldwide market' (Roco, Hersam et al. 2011: 1). A seminal proposal for nanotechnology based on an American government report from the National Science and Technology Council, focussed as was identified in the title of the report, on shaping the world atom by atom (Amato 2000). The report builds on predictions for the future and unearths some of the pending discourse as to a nanotechnological interpretation of life.

The shaping of the world, implicated in the title of the report implies a reshaping of nature via nanotechnology. The suggested re-shaping being a form of molecular self-assembly that seeks to use concepts of 'supramolecular chemistry is the designed chemistry of the intermolecular bond. In other words molecules are synthesized that will interact with others of the same or of a different type, using weak non-covalent intermolecular forces. This leads to the spontaneous linking together of individual building blocks into complex molecular assemblies of a planned type' (Bishop).

In effect this is nano assembly, or a 'bottom-up' approach, that is, the construction of a supramolecular chemistry by the assembly of nanoscopic particles, or even atoms and molecules. The desire for nanotechnology is to invent new truths and then create devices that can utilize 'forces of nature'.

The report to the National Science and Technology Council demonstrates an attitude to materiality and life stating

> if you were to deconstruct a human body into its most basic ingredients, you'd get a little tank each of oxygen, hydrogen, and nitrogen. There would be piddling piles of carbon, calcium, and salt. You'd squint at pinches of sulfur, phosphorus, iron, and magnesium, and tiny dots of 20 or so other chemical elements. Total street value: not much.
>
> (Amato 2000)

The 'not much' takes the focus from the atomic and put its back on the coalescence of all these materials to have the scale and temperament to contain life. As chemical compounds we are of little value and have no real call on life; we become merely commodities of a mechanistic science.

The Amato report referenced in Chapter 1 quotes Smalley and Stormer whose use of language is striking. For example Smalley suggests that nanotechnology is 'the builder's final frontier' and Stormer sees it as, 'the ultimate toy box of nature'. These statements reinforce the idea of nanotechnology's potentially, 'limitless ability to create new things' (Amato 2000). Similarly Mirkin's language echoes these views, he suggests that nanotechnology is 'inspiration from life to create new forms of matter … it's a real example of man over nature' (Steenhuysen 2008). These claims and the visual image on the cover of the National Science and Technology Council report are analysed by Alfred Nordmann. He suggests the cover image makes reference to the conquest of outer space being matched by the conquest of inner space. The foreground of the cover shows a floating nano surface image travelling through space towards the earth. The planet Earth in the background reflects the macroworld while the scan of atoms deals with the micro. The liminal space between the two worlds is made clear, and the scales represented challenge a human viewing of the scene. The human becomes disembodied and dematerialized to enable the shift between the different worlds of research (Nordmann 2004).

The various quotes used in the US government report demonstrate a quantitative materialistic and non-sensorial reconfiguration of our position in the world. What is expressed here is a perspectival imposition in seeing the world. The human being is positioned as having ownership of all matter and therefore the rights to play in the reconstruction of nature. In addition, what nature is becomes confused and consumes any definition of life. In Nordmann's critique the nanotechnological rhetoric is the vision of scientists as space pioneers with ethical and moral stances boldly going where no man has gone.

The developmental process of life has now moved into a post-human, and post-biological discourse of human and machine with the evolution of the 'molecule to sign' (Hoffmeyer 1994). We are also confronted with another transformation as 'moist media' is reinvested with molecular digital signs that exist at the nano level. The analogically coded sign of the cell and the digital code of the DNA are the main distinction being made through the language of nanotechnology.

The Action Group on Erosion, Technology and Concentration (ETC Group), who are dedicated to the conservation and sustainable advancement of cultural and ecological diversity and human rights, are dealing with socially responsible developments in technologies. They explain:

> At the nano scale, where objects are measured in billionths of meters, the distinction between living and non-living blurs … [H]uman-made nanomachines that are powered by materials taken from living cells are a reality today. It won't be long before more and more of the cells' working parts are drafted into the service of human-made nanomachines. As the merging of living-nano and non-living nano becomes more common, the idea of *self-replicating* nanomachines seems less and less like a "futurist's daydream"'
>
> (ETC Group 2003: 2–5)

This blur between living, non-living and vibrational matter can be seen to be explored in the artworks *Nanoessence* and *Blue Morph*. The works draw together some of the intriguing questions posed by the ETC Group. As nanotechnological experimentation develops concepts of rebuilding the world atom by atom it exposes such questions of where life might exists? We can look at sonic 'Elementary vibration' in the works of Susan Alexjander, David Deamer, James Gimzewski, Andrew Pelling, Anne Niemetz, Adam Zaretsky, Oron Catts and Ionat Zurr as having clear relationship with a haptic redefinition of a causal challenge to our sense of self in respect to our self as matter, and in turn all matter. The next section will examine a range of works that aim to engage with vibration at an atomic level.

## Sonic presentation

The translation of science through sonic presentation is an artistic tactic in which an invisible world is being translated to open a dialogue with the infinitely small. The following are two examples of work that focusses on sound.

In 1988 a sound artwork was produced by composer Susan Alexjander. She worked with cell biologist David Deamer at the University of California after he released *DNA Suite* and *DNA Music*.[6] These sound recordings were based on data mappings of sequences of adenine, thymine, guanine and cytosine (DNA bases) gathered along the helix. Alexjander and Deamer worked together on a project that created compositions based purely on the chemical composition of the four DNA bases. The molecular vibrations of the bases were measured using an infrared spectrophotometer. The process involved exposing each DNA base to infrared light, measuring the absorbance of their wavelengths and then transcribing light into sound with the help of a Yamaha synthesizer. The compositions were released on tape cassette titled *Earthday* in 1990. Subsequently they were released on CD, titled *Sequencia,* as three original performances using traditional instruments and electronic keyboards entitled *Eikos, Sequencia* and *Pataphysical Thymine* (Alexjander 2007).

In 2002, Adam Zaretsky, a bio-artist working in a laboratory at the Massachusetts Institute of Technology, discovered, what he called the 'Humperdinck Effect', in which the vibrations of loud, really awful lounge music applied for 48 hours, spurred antibiotic production in *E. coli* bacteria. Subsequently Zaretsky met with Oron Catts and Ionat Zurr of the *Tissue Culture and Art Project* and collaborated on *The Pig Wings Project* in developing the *Dynamic Seeding Musical Bioreactor*. The project involved experimenting with submitting bone marrow cells, as they were being cultured, to audio vibrations (Figure 1). The work explored the use of irregular vibrations to distribute the cells throughout the scaffolding and used audio vibrations taken from a Napster[7] search for songs with 'pig' in the title.

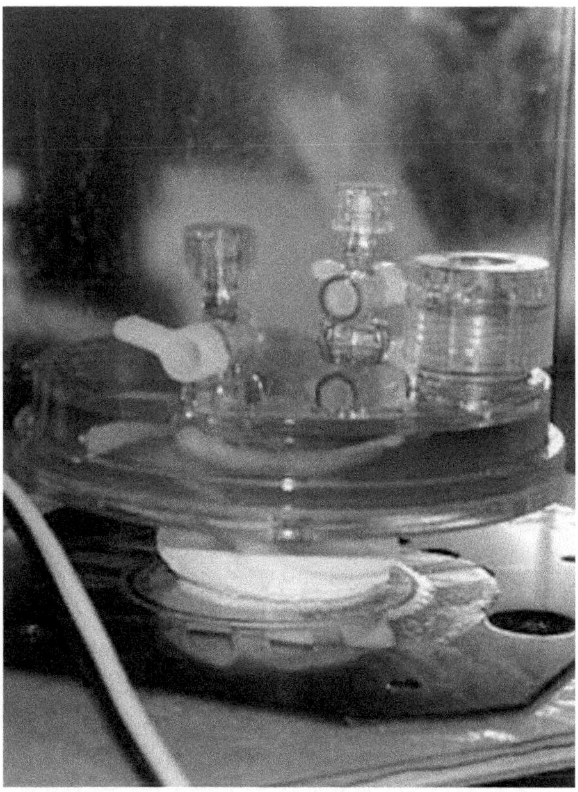

**Figure 1:** Dynamic seeding of pig Mesenchenal stem cells onto/into degradable polymer scaffold in shape of wings, using a vibrating speaker that played pig songs. By Ionat Zurr (2002).

The songs were, played to the cells by placing a speaker underneath the Petri dish, and included *War Pigs* by Black Sabbath; *Fascist Pig* by Suicidal Tendencies; *Da Killing of Da Pigs* by Da Yoopers; *Chokin' this Pig* by Eminem; *Squeal Like a Pig* by the Reverend Horton Heat; *Filth Pig* by Ministry; *American Pigs* by the Angry Samoans; *British Pigs-The Price of Royalty* by One Life Choir; *PigInCheez* by Aphex Twin; *Blue Christmas* by Porky Pig; and *Pigs on the Wing* by Pink Floyd.

The *Pigs Wings* research project may seem to be counter intuitive as bone marrow vibration is at the core of stimulating and communicating with matter at a nano level. However, in this project it was the rhythms, and patterns of music that could communicate below the cellular level directly with the cells' atomic structure.

**Vibrant cells**

The work of Zaretsky, Catts and Zurr occurred simultaneously to a project being developed between James Gimzewski and Andrew Pelling (2002). Gimzewski and Pelling had discovered that atomic sounds were emitted by yeast cells. They termed the study of this phenomenon: 'sonocytology' (Pelling, Sehati et al. 2004). In what appears to be a complete reversal of the *Pigs Wing* project Gimzewski and Pelling recorded atomic vibration and translated it through a computer program capable of converting the nano-scale vibration into an audio file.

At this time a range of exhibitions were being staged to bring nanotechnology before a broader audience and to visualize what is taking place at a nano level, including in 2004 a nano exhibition at the LA County Museum of Art. The exhibition showed the results of collaborations between N. Katherine Hayles, Robert Sain, Victoria Vesna and James Gimzewski. Other collaborations included artist Anne Niemetz and a researcher Andrew Pelling who used the AFM to collaborate on one of the works within the exhibition titled *The Dark Side of the Cell* (Figure 2).

The work premiered on 2 June 2004 and was a sonic immersive nano installation that for the first time utilized cell sonics. Nano-scale imagery of the cells was projected, accompanied by sonograms of the cells, onto sculptural elements based on the inner structure of the cells. Niemetz described the AFM as a new type of musical instrument due to the machines ability to tune into the atomic world (Niemetz 2004). The sounds created by Niemetz and Pelling in *The Dark Side of the Cell* comprised five separate movements using the vibrational sounds taken from the yeast cells at various temperatures. Pelling explains and comments on the process of recording atoms with the AFM:

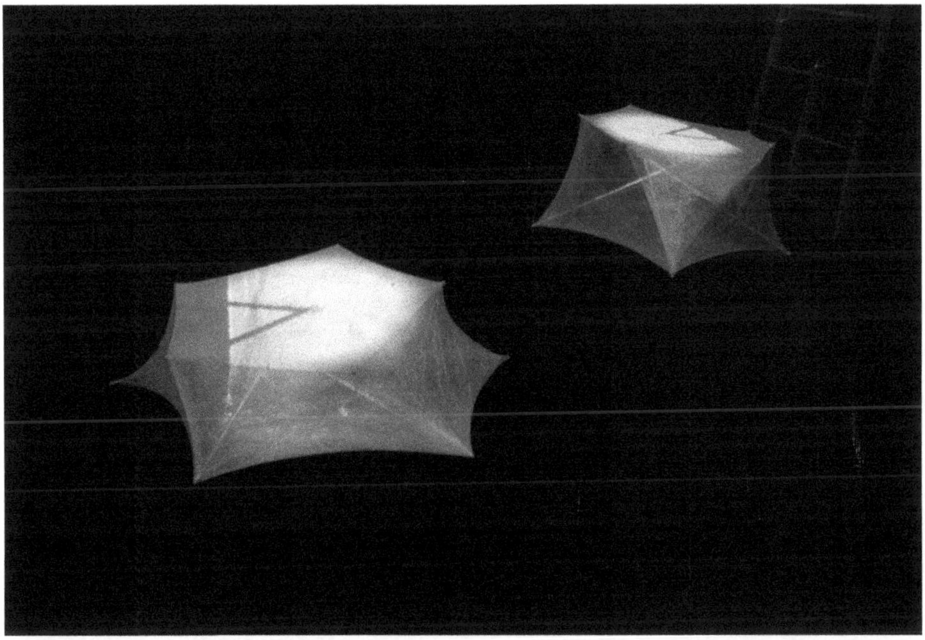

**Figure 2:** Anne Niemetz and Andrew Pelling, *The Dark Side of the Cell* (2004). Los Angeles County Museum of Art.

> [T]he tip is held stationary over a surface that is vibrating or moving, the tip will bend and follow these motions. This fact opens up a door for a whole new set of experiments. The AFM can now be used not only as an imaging tool, but also as an ultra-sensitive, high-resolution motion detector.
>
> (Pelling, Sehati et al. 2004)

*The Dark Side of the Cell* art work explores how sonocytology and the AFM are potentially a new form of listening to the sonic vibrations and movement at the heart of the immaterial world. The atoms recorded create rhythms and patterns that we are able to recognize and interpret in relation to our own innate material vibrations.

## Blue Morph

Vesna's and Gimzewski's *Blue Morph* (2007) is an interactive installation that uses nano-scale images and sounds derived from the metamorphosis of a caterpillar into a butterfly. The research for Blue Morph examined the mechano-structural properties of the Blue Morpho butterfly (*Morpho peleides*) to develop an understanding of its

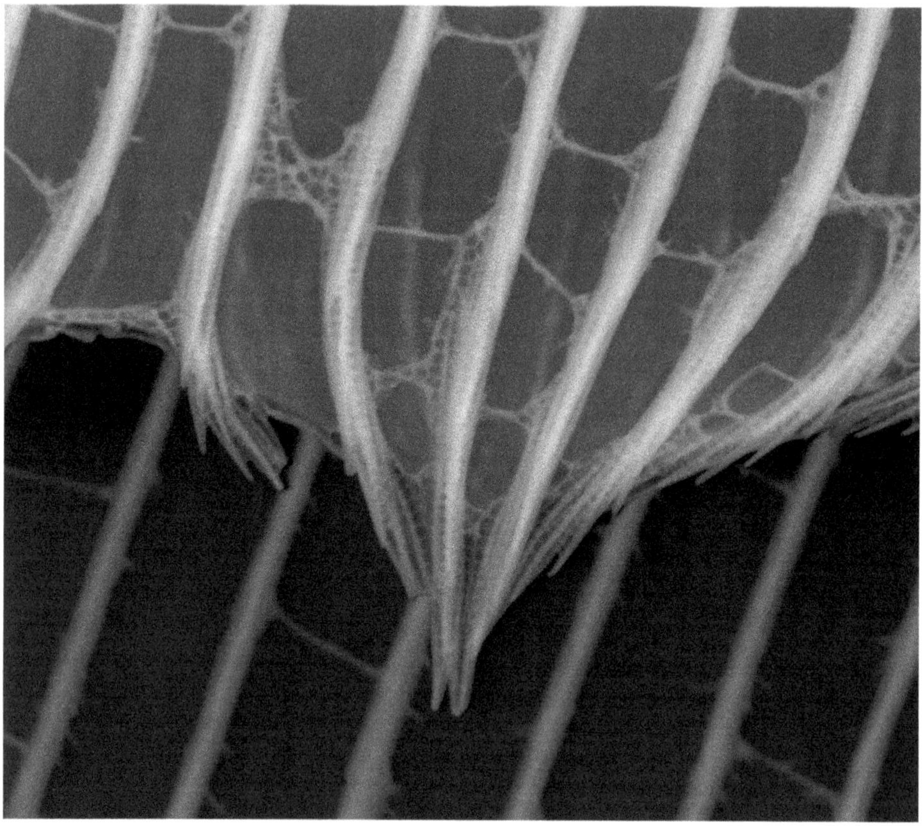

**Figure 3:** SEM Butterfly wing cell image at FEI corp, Oregon. By James Gimzewski (2007).

optical properties. Along with the optical properties scanned with an Scanning Electron Microscope (SEM) (Figure 3) was the pupa, which was also scanned with the AFM using the technological process of sonocytology.[8] The machinic data generated by the AFM in force spectroscopy mode were sonic vibrations of the caterpillar's membrane. The data demonstrated that the cellular transformation, taking place within the pupa, was happening in surges of vibrations.

The embodied installation has existed in various incarnations from its first iteration. In one version a user, wearing a simulated floating weather balloon that has been fashioned at the narrow end to fit on to the head, sits on a prepared surface in a meditative state of stillness and silence. The weather balloon acts as a metaphorical antenna linking the vibration of life to a larger cosmic vibration as the user experiences the fundamental vibration of life, that is the pupa morphing and evolving from its material beginnings. It allows the viewer to be in-touch with the primordial palpable experience. To hear and feel the signification of energy at an atomic level illustrates the workings of evolution, emergence and becoming.

The work poses questions of duration. The intermittent surges of energy, discovered in the scientific analysis of the recordings made by the AFM from the pupa, are inherent below the cellular level, in the matter. The work does not shy away from chasing scientific discoveries, but is led by the data into intuitively constructing a vision of the interconnecting patterns of rhythm and vibrations that pose questions of life. The surges of vibration inherent within the pupa align with a shift in harmonics, the creation of milieus.

## Nanoessence

The artistic relationship between the concept of life and vibrational vital matter is explored in the interactive audio-visual installation work *Nanoessence*. In this installation the viewer interfaces through his or her own breath, with a visual and sonic presentation of a metaphorical spatial envelope between life and death. Breath plays a quintessential role conceptually with its strong link to the biblical inception of life.

The project as with the previous work *Blue Morph* attempts to maintain a high degree of authenticity to engage the viewer in a sensorial qualitative experience of quantitative data pertaining to life.

The *Nanoessence* project aimed at examining life at a sub-cellular level and re-examining matter, space and scale within the human context. A single HaCat skin cell was analysed with an AFM to explore comparisons between life and death at the nano level (Figure 4). The work explored a humanistic discourse concerning concepts of what constitutes life being challenged by nanotechnological research.

The project maintains a differentiation between science and art by accepting a variance in the rhetoric used to describe each of the various sets of conditions. The work explores the difference between quantitative text that illustrates scientific research and art that challenges and aestheticizes assumptions based on substantive qualitative text.

### *Immortality and the experiments*

The interconnected relationships between the HaCat cell, the AFM and the breath interface together constitute the context of the *Nanoessence* project. The cells chosen for the *Nanoessence* project were HaCat cells. Boukamp explains: 'HaCat is the first permanent epithelial cell line from adult human skin that exhibits normal differentiation and provides a promising tool for studying regulation of keratinization

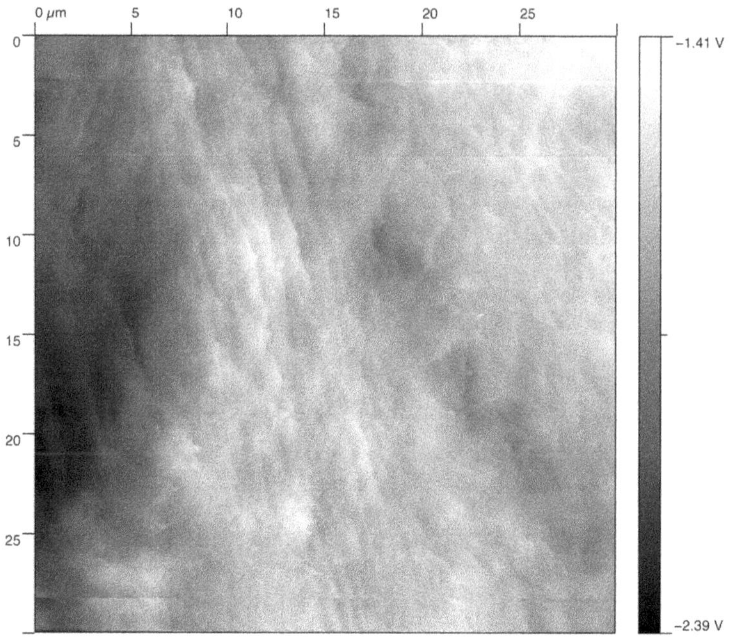


**Figure 4:** HaCat skin cell image from an Atomic Force Microscope. Part of the *Nanoessence* project by Paul Thomas (2008).

in human cells' (Boukamp, Petrussevska et al. 1988).[9] Importantly, these cells 'although immortal, have largely retained their capacity to reconstitute a well structured epidermis after transplantation in vivo' (Nykypanchuk, Maye et al. 2008). The HaCat like other cells have an engineered immortality and therefore demonstrate the potential of endless cloning of a single cell.

The creation of immortality was usually the prerogative of gods and not humans. The development of an immortal living cell line has a relationship with the creation of life as in the book of Genesis. The tree of life was planted by God in midst of the Garden of Eden (Paradise) where the fruit could bestow everlasting life to the person who ate it. It was due to a moral failure that the opportunity for immortality was lost, whereas now scientific cellular immortality can be constructed in a Petri dish.

***Breath***

The rationale for the use of breath and moisture as the interface for the *Nanoessence* project is contextualized through the reading of King James (2000) bible Genesis 2: 7. In this section of the bible, dust particles were gathered together by God to create man. God then having created a lifeless material shape transmitted life by breathing into figures sculptured nostrils, 'And the Lord God formed man of the dust of the ground, and breathed into his nostrils the breath of life; and man became a living soul.'[10] The act of transmitting life is relative in this context as it suggests that life cannot be manufactured but needs to be transmitted. The haptic quality of breath as a material construct amplifies a form of touch at a nano level. The breath acts as a paradox of transmission between the material and immaterial, touching upon properties of post-humanism as opposed to being human.

In the Qu'ran it is stated that 'He directed Himself to the heaven, and it was a vapor, so He said to it and to the earth: "Come both of you, willingly or unwillingly." They both said: "We come willingly".'[11]

Mohamad Jawad Chirri stated in an online interview with Wilson H. Guertin when speaking about this extract from the Qu'ran that 'the quoted verse indicates that the vapor or what constitutes the vapor of molecules and atoms was the first material thing that existed in this world'. Pedro C. Marijuán in his article 'Bionformation: untangling the networks of life' suggests: 'Water is the starting point. Necessarily, it has to be considered the first biomolecule, the primordial one which has "selected" all the other molecules participating in the evolutionary plays of life' (Marijuán 2002).

In *Nanoessence* the HaCat cell, the AFM and breath linked with vapour, demonstrate a fundamental symbiotic relationship in the instigation of life at a nano level. The dust particles, vapour and breath are fundamental concepts and materials in the installation's investigation of the creation of life. The dust particles are revealed like 'steam thrown into the air is nearly all condensed into little drops which fall back, and this condensation and this fall represent simply the of loss something (*sic*), and interruption, a deficit' (Bergson 1911: 26). The steam's vapour thrown into the air reflects on atoms that are falling endlessly in the void, they represent simply the loss of being. The breath becomes the unpredictable element of the swerve that is at the start of life.

### *AFM*

The link between breath and the AFM in the work *Nanoessence* is made through the use of the cantilevers that use vibration to reveal proximity at a nano level, by translating recorded data into topographical visualizations and sonifications. The vibrancy of matter is transferred to the mechanism of the AFM's cantilever that is designed for recording atoms. The AFM uses sonic frequency to control the cantilevers as they seek out atomic patterns, rhythms and molecular vibrations. The AFM designed by humans to make the invisible world of atoms visible reveals humans as 'vibrant matter'. Therefore, this suggests that if we see all matter as consisting of small amounts of latent material chemical compounds in the process of becoming, there has to be a shift in how we currently consciously discriminate and make distinctions between forms of matter.

***Bio to nano methodology***

At this stage, I would like to give a fuller account of the structure of the *Nanoessence* research project and reflect on the experience to make intuitive insights based on the experience of undertaking the project. The *Nanoessence* research commenced in 2008 and was based on data and information gathered as part of a residency at SymbioticA, Centre of Excellence in Biological Arts, University of Western Australia and the Nanochemistry Research Institute (NRI), Curtin University of Technology.

The project used HaCat human skin cells to explore life and death at a nano-scale. The HaCat cells that had been provided for use in the project were 'full thickness adult human body skin'. They were 'obtained by surgical excision … from the distant periphery of a melanoma located on the upper half of the back (not extensively sun-exposed) of a 62-year-old male patient. The histology of the epidermis from the skin specimen obtained in a second, "safety" operation at the primary melanoma site showed no apparent anomalies' (Boukamp, Petrussevska et al. 1988: 78).

The HaCat skin cells, cultured with the assistance of Ionat Zurr from SymbioticA, were scanned by the AFM in tapping and force spectroscopy mode (as discussed in Chapter 2) to determine comparative topographies and atomic vibration as the cantilever tip scans the cell.

Mica substrates, scaled to fit the AFM scanning Petri dish, were sterilized and given a coating of collagen (extracellular matrix) to provide structural support and enable the skin cells to attach more securely to the surface. The samples coated with collagen were put into an incubator for 1 hour to allow the collagen to cross-link with the mica. The excess collagen was then washed away and the samples were ready to receive the cells.

The cells were extracted from the tissue culture flask by first removing the nutrient solution (which contained serum) and then using trypsin, which is a digestive enzyme that causes the cells to detach themselves from the surface on which they are growing. The cells were then added to the mica samples and placed in a 37°C/5%$CO_2$ incubator. Within 2 days the cells were confluent on the six mica substrates and ready to be transported to the Nanochemistry Research Institute at Curtin University.

The samples were collected from SymbioticA at 10 a.m. and placed in a transportable incubator. The living HaCat skin cells were unpacked from the transportable incubator at 11 a.m. at the NRI and one mica substrate sample was placed under a watertight mounting (Figure 5) which was to be to be filled with serum and then inserted into the AFM.

The AFM was set up in tapping mode and initial scanning commenced at 11.30 a.m. when a living HaCat skin cell was raster scanned. In tapping mode a piezoelectric actuator is used to control the height of the cantilever as it scans above the skin cell.

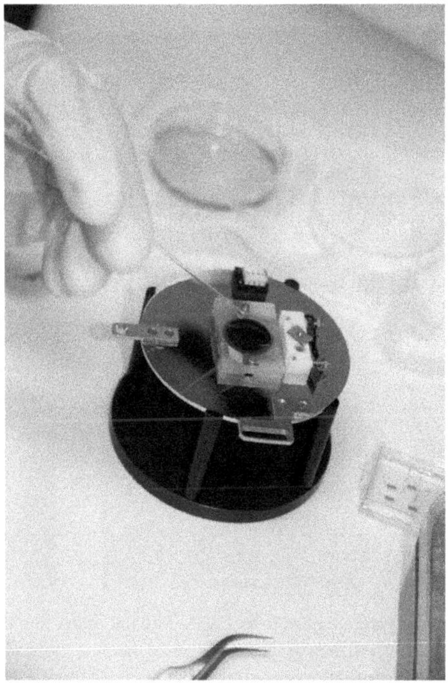

**Figure 5:** The mica substrate is placed on holder to be inserted into the AFM. The rubber circlip is placed under AFM polystyrene Petri dish. The Petri dish is assembled before the serum is placed back over the cells.

This mode allows for a more sensitive recording and is less likely to damage the cell structures. The cantilever makes intermittent contact as it is lowered to touch the surface of the skin cell and oscillates up and down in a tapping motion created by a resonance frequency. The oscillations creating the tapping motion are between 100 nm and 200 nm and so the space between the surface and the cantilever is infinitesimal.

An area of the specimen measuring 30 × 30 microns was scanned and the images obtained from the cantilever's motions were recorded via a laser beam deflection from the cantilever tip onto the photodiode (Figure 6). The AFM cantilever explores the surface topographies gathering data to produce a range of formats including, topography, friction, phase, deflection and amplitude, and to create images.[12] *Nanoessence* employed phase imaging, which is described as a

> powerful extension of TappingMode™. The AFM provides nanometer-scale information often not revealed by other SPM techniques. For example, phase imaging goes beyond

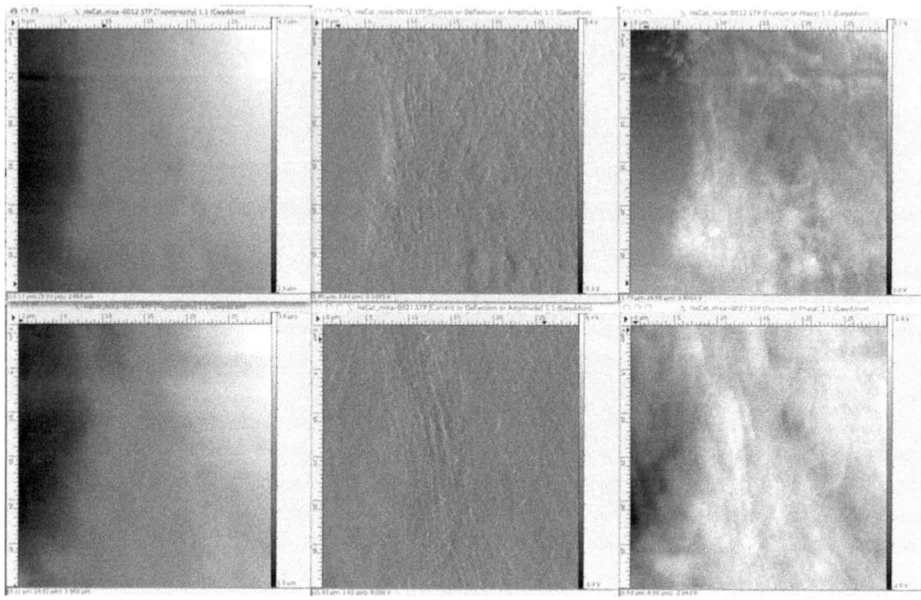

**Figure 6:** Three modes of image representation from the AFM data. Image shows three modes of image representation from the AFM data; (Left) topography, (middle) deflection or amplitude, (right) friction or phase using Gwyddion software. The images represent scans of living HaCat cell (top row) and scans of a dead HaCat cell (bottom row). The cells were scanned for the duration of 2 hours.

> topographical mapping and can detect variations in composition, adhesion, friction, viscoelasticity, and other properties. The resolution of phase imaging is comparable to the full resolution of TappingMode and can act as a real-time contrast enhancement technique.
>
> (Benatar 1997)

The initial set of experiments (outlined above) was repeated under the same conditions the following day with a scale adjustment, reduced down to 2.5 × 2.5 microns. The scans were also produced over a 2-hour period that allowed for the HaCat cells to expire.

During the second day a different AFM in the NRI was also used, this time in force spectroscopy mode. This mode does not scan the surface of the cell but determines the atomic vibration as the tip touches and then rests on the surface of the cell. Thus, in force spectroscopy mode data to ascertain the reaction of ether to a group of atoms of the HaCat cells were gathered. The living skin cell was initially sampled *in vitro* and the atomic vibration was recorded for a period of five seconds. The serum immersing the skin cell was injected with ether while the scanning process was taking place and the cell was recorded for a further 5 seconds.

A meeting was set up with Dr Peter Hinterdorfer, associate professor at the Institute for Biophysics, Johannes Kepler University of Linz, to discuss the scientific reading of my practical research data from the NRI. Hinterdorfer examined the data from both experiments and suggested that when the skin cell was recorded in phase mode the head of the cantilever was tapping and making contact with the skin and was recording both the attraction and resistance. He pointed out that in the original experiments the surface of the skin cell had shrunk and tightened revealing the internal structural scaffolding and lines of communication within the cell. Thus the visual images recorded in phase mode revealed the HaCat cell's structural supports that carry information throughout the cell.

***Nanoessence installation***

The *Nanoessence* project attempts to explore some of the issues relating to the creation of life, through the form of the installation and its interface. The role of the artwork is to reiterate and bring together some of the debate being processed through a post-humanist discourse. In *Nanoessence*, the viewers breathe onto a rapid prototype model of the first AFM HaCat cell scan. Data from the breath and its vapour is used to generate genetic algorithmically constructed cellular automatons.

A series of metaphorical landscapes were created using the data from the HaCat cells recorded by the AFM. The landscapes consisted of layers of images that were a gradation of living cells through to dead. The individual data scans when layered on top of one another (like filo pastry) represent a landscape, that when entered virtually reveals a space between life and death. The data projector displays nano level topographies of skin cells that represent the envelope of space in which primordial cellular automaton structures emerge. The structures arise from algorithms developed by Kevin Raxworthy that define a series of problems relative to becoming. The algorithm is stimulated by information gathered from sensors that respond to the viewer's breath. The power of the breath and moisture reading create and stimulate parameters for the essence of life to grow.

Sonic topographic textures in *Nanoessence* are created from the analysis of data recorded by the AFM in force spectroscopy mode. The vibrations of the HaCat cell atoms are scanned initially *in vitro* and then a second time with ether injected into the serum. The resulting comparative analysis of the data is converted into sound files. The installation presents an auditory work and a spatially orientated haptic topographic sensation based on sonic vibrations that occur at the nano level (Figure 7). The auditory work sonically maps out the topographical relationships between the changing landscapes.

The aim of *Nanoessence* is to engage not only the visual, but also other senses in an understanding of infinite smallness, and create a psychological shift in the viewer

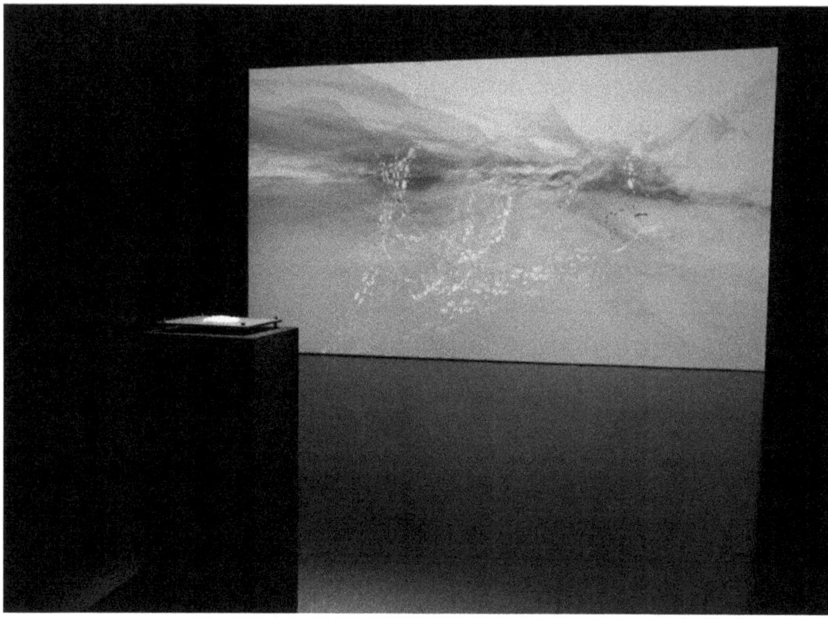

**Figure 7:** Paul Thomas, *Nanoessence* (2010). Installation view. John Curtin Gallery.

by using breath as an interfacing device that re-orientates the viewer to identify both sonic and visual differences between a living and dead single skin cell.

The lengthy description of the process presented here is to demonstrate the relationship between research and what is needed to achieve a comprehensive understanding of pseudo-scientific research. Getting a balance between the levels of science and art in a project is difficult; it raises questions about what are the levels of scientific understanding that should be explored by the artist in the pursuit of the artwork. The science in the artwork *Nanoessence* created a structure which triggered questions to me as the artist about being. This offered a series of challenges that intuitively related to the commitment already made to my ongoing practice in this area. Making connections between the sciences and the humanities generates a new dialogue, venues and audiences which allow for a rethinking of our understanding of the world.

## Nano futures

The installation of *Dark Side of the Cell* explored aspects of rearticulating scientific data through a sonification process to create new forms of rhythm and pattern recognition. The sounds, born out of the yeast cell being exposed to shifts in temperature, were translated into patterns that reflect life within matter. *Blue Morph* presented data in a contemplative environment to align the viewer with the reverberation of becoming. In *Nanoessence* the transmission of the viewer's breath stimulates, transforms and sustains algorithmically generated cellular automatons, creating a representation of artificial life. It places the viewer at a critical point in the discernment of a humanist comprehension of matter and life. The works point to an understanding of the material world through an exposure to the core of our own intrinsic embedded vibration.

Questions about life exist as language, a code that inherently implies that the object of life is also held within that code. Transmission of the code, as in the case of breath, is via a network where it can be distributed and reengineered to recreate an 'object'. Similarly to associate code with thought is to suggest that the thought of life inherently holds the material construction of life. For the idea of life to come from life there must inherently be life in that code. Code is analogous to swarms of atoms creating meanings from a machinic process that demands recognition of its patterns to exist.

The material world that is constructed from the swarm of atoms that exist to form a cell or an organ are all 'born out of chaos' (Deleuze and Guattari 1987: 314). This chaos is primordial; it is pattern, vibration and rhythm that construct the boundaries of matter and meaning. This meaning is part of our ability to determine patterns. Hayles suggests: '[If]pattern is the realization of a certain set of possibilities, randomness is the much, much larger set of everything else, from phenomena that cannot be rendered coherent by a given system's organization to those it cannot perceive at all' (Hayles 1999).

Hayles' post-human subject is a construct born out of chaos, a pattern within the information feedback loop. The post-human feedback loop is based on a relationship with technology. Hayles uses the example of the blind man's cane to illustrate the feedback loop, suggesting that in a cybernetic relationship the cane is part of the human. Therefore the realization of pattern creates a set of possibilities for the human to be part machine. This argument garners further support from the recognition of the symbiotic relationship between technology and the seer, which perceives technology as a reflection of the seer. Merleau-Ponty states:

> [S]ince the seer is caught up in what he sees, it is still himself he sees: the object is the mirror in that the seer and the visible reciprocate one another and we no longer know which sees and which is seen.
>
> (Merleau-Ponty 1968: 139)

Here the human and machine become one as the seer and seen coalesce the processes into a symbiosis.

The swarm of atoms then is to be constructed in many different configurations based on the individual's ability to define patterns while also being a pre-existing example of a recognized pattern. The formation of a recognizable pattern that is durable and consistent allows at that point for the pattern to be determined, understood and communicated. The feedback loop of the blind man's cane is also the story of the totality of patterns sensed and realized at a nano level through the touch of the AFM's cantilever.

Nanotechnology allows for installations to occur at the level of the construction and transmission of life's patterns, meanings and rhythms; not in the material sense of reshaping nature atom by atom but in the complex interrogation of the atom's role in becoming life.

## Notes

1 Kant in 'Critique of Judgment' distinguished desire and will as concepts, and refers to matter as being lifeless and mechanistic (but rather, as with lifeless matter, by mechanism, and, as with the lower animals, by instinct), Kant, I. (1987), *Critique of judgment/Immanuel Kant,* translated, with an introduction, by Werner S. Pluhar; with a foreword by Mary Gregor, Indianapolis, Ind.: Hackett Pub.

2 Hans Driesch was a biologist and philosopher whose experiments with embryology informed his theory of entelechy (a life giving force). His work with organisms revealed to him that this life-giving force was an 'invisible presence' that could not be performed by a machine. Bennett, J. (2010), *Vibrant Matter: A Polical Ecology of Things*, Durham: Duke Univeristy Press.

3 Michael Vaughan suggests that Bergson's élan vital does not predict an immaterial force but, 'on the contrary, designates the vitality of matter itself, its organization, its growth, its indeterminacy, unpredictability and creativity, and this is inaccessible to mechanism in principle, not merely in fact'. Vaughan, M. (2007), 'Introduction: Henri Bergson's "Creative Evolution"', *SubStance*, 36: 114, pp. 7–24.

4 The swarm behaviour is exhibited by animals of similar size which aggregate together, 'the concept of the swarm is predominantly a biological concept – its roots are to be found in the study of ethology (animal behavior), and can be traced back to nineteenth century natural science'. Thacker, E. (2004, 2007), 'Networks, swarms, multitudes', from http://www.ctheory.net/articles.aspx?id=423.

5 Nanotechnology takes the words bottom-up literally to mean building things from the bottom-up, with atomic precision.

6 David Deamer and Susan Alexjander, 'DNAMusic'/'DNASuite', CD (Aptos, CA: nodate), see: Alexjander on http://www.oursounduniverse. com/music_samples. html. Accessed 20 May 2012.

7 Napster was a name given to a pioneering peer-to-peer file sharing Internet service of audio files, music, encoded in MP3 format. The company was shut down by a court order for copyright infringement.

8 '"Sonocytology", the term Prof. Gimzewski suggests for this cutting edge field of cell sound study, represents a new realm of challenge and potential for scientists, artists, and in particular for musicians.' Niemetz, A. (2004), *Singing Cells, Art, Science and the Noise in Between,* Master of Fine Art, University College Los Angeles.

9 The apparently immortalized but highly differentiated cell line was named HaCat to indicate the origin and initial growth conditions Boukamp, P., Petrussevska, R. T. et al. (1988), 'Normal keratinization in a spontaneously immortalized aneuploid human keratinocyte cell line', *Journal of Cell Biology*, 106: 3, pp. 761–771.

10 http://king-james-bible.classic-literature.co.uk/genesis/ebook-page-02.asp, accessed 24/03/21012.

11 http://quran.com/41/11 and http://www.al-islam.org/inquiries/4.html, accessed 24/03/21012.

12 The (*a*OCT) 'directs a light beam perpendicular to the catheter-distance between the probe head and the air–tissue interface of the airway wall is determined from the reflected light using a low-coherence optical interferometer. Probe rotates at 1.25 Hz to capture quantitative cross-sectional images of the upper airway' Wang, J., Tetlow. G. A., Lucey, A. D., Armstrong, J. J., Leigh, M. S., Paduch, A., Sampson, D. D., Walsh, J. H., Eastwood, P. R., Hillman, D. R. and Harrison, S. (2006), 'Dynamics of the human upper airway: on the development of a three-dimensional computational model', *IFMBE Proceedings World Congress on Medical Physics and Biomechanical Engineering,* vol. 14.

# Chapter 4

## Matter, Measurement and Light

In all matter there is the possibility of being and nonbeing. Matter that has already been considered as not being could now demonstrate a more latent potential than being. Being, then, is the major concern in defining this chapter's area of focus. The concept of 'nonbeing'[1] pervades our awareness of matter. For example, the saying 'as dead as a door nail' is analogous to matter having no life, of being finished with and unable to be reused. The question of vitalism and vibration as a perception of matter developed previously is being seen by nanotechnology as a quantifiable resource. This chapter intends to challenge and critique the concept of ownership of the material world by the theories and artworks explored by the artists.

The creation of quantifiable data as a determination of life is a blindness at the heart of our objectification of the material world. The very nature of this thinking is challenged by the genealogy given to us by Lucretius, Bergson and Serres in the form of unpredictability, uncertainty and immeasurability. What Lucretius gives us is an insight into a measured state not as a recordable certainty, but as an unpredictable potential in the world. Serres when defining the clinamen says: 'it is the minimal angle of formation of a vortex, appearing by chance in a laminar flow' (Serres 2000: 6).

The constant shifting, unpredictability, uncertainty and immeasurability[2] of these various modalities contest scientific theories that claim to deal in facts. However, the idea of contestable theories is currently being illustrated by the debate surrounding the measurement of the speed of light. According to news reports from European Organization for Nuclear Research (CERN), neutrinos (sub-atomic particles) were sent through rock to Italy some 730 kilometres away. They arrived 60 billionths of a second faster than the speed of light (Evans 2011). Current research theorizes that these measurements demonstrate that there is likelihood that the speed of light is not a constant. This is a test of human consciousness and its ability to comprehend and evaluate this infinitesimal measurement. Furthermore, we are confronted by the exposure of a falsity in what has always been thought of as a quantifiable fact.

The recent potential discovery of the Higgs boson at CERN in 2012 as an instrumental particle in the development of mass within the universe is now, through media coverage, becoming part of our human expanding consciousness. The Higgs boson is part of the phase transition[3] in the process of defining what holds this universe together. As Lucretius states, atoms unpredictably swerve and in so doing obtain mass and form. We find in the current understanding of the Higgs boson that some particles get a mass and some don't, the reason for this, like all other reasons of existence, is dependent on a number of probabilities. The one thing that is known is that something needs to happen, to change; you need to change some of the symmetries that make up the universe of atoms falling in the void to obtain individuality. If at the dawn of time atoms were equally and symmetrically dispersed throughout the universe there needed to be a perturbation that would allow for the 'breaking of symmetries [which] is absolutely vital for our existence' (Copeland, Bowley et al. 2011).

## Quantifiable measurement

Quantifiable measurements are part of our ongoing modern understanding and assessment of the material world that, according to Lissa Roberts in *The Death of the Sensuous Chemist*, separates the human from these experiments. Roberts' enquiry into the shift away from use of the senses in scientific investigation in the late eighteenth century states: 'chemists increasingly subordinated their bodies to the material technology of their laboratories and began erasing the presence of direct sensory evidence from the public records of their discipline's literary and social technologies' (Roberts 2004: 507).

Roberts explores how the demands for quantifiable, predictable and transportable data sets shaped our comprehension of nature and observes that there is a growing reliance on data to understand contemporary living. What Roberts is demonstrating is a moment in time when there is a departure from human experience and a transfer over to a mechanistic understanding of materiality. She highlights the beginning of what has become, in our digital age, an ever-increasing amount of quantifiable data that needs to be decoded and interpreted by the machines that created them. Roberts explains that philosophical discourse in the eighteenth century suggested that

> mathematical physics offered a superficial account of nature, achieved without first-hand experience of natural processes and at the expense of obliterating the knowledge of nature's heterogeneous richness that such experience affords. Because common artisans manipulated nature on a daily basis, they knew more about its workings than did mathematical physicists. But, proceeding without metered reflection, artisans could not systematize and teach the wisdom that lay behind their skills.
>
> (Roberts 2004: 503–504)

Roberts questions that moment in time of the death of the sensuous chemist, she highlights how the smell, feel, taste of a substance became no longer relevant. Instead there is the need for categories and systems where information is objectified and allows for data to be substituted for experience instead of 'arid reason and arts intuition' (Roberts 2004: 504).

The substitution of experience, when the data became quantitative and not qualitative, can be aligned with 'Zeno's paradoxes';[4] a series of hypothetical problems devised by the Greek philosopher. Three of the paradoxes are, 'Achilles and the tortoise', 'the dichotomy paradox' and 'the arrow paradox'. They dealt with measurement and motion and are relevant to this chapter. They highlight the dilemma relating to the use of measurement, particularly with regard to how it is interpreted and of what relevance it has to being human.

## Paradox

Zeno's 'arrow paradox' relates to measuring each position in time in an arrow's line of flight. Zeno assumes that it travels no distance during the moment that 'it occupies an equal space' for the whole instant. According to this argument the entire period of its motion encompasses only instants, all of which contain an arrow at rest. Therefore, Zeno concludes, the arrow cannot be moving.

We can examine this concept through the example of timing athletes in a 100-metre sprint. An electronic timekeeper can document each runner's time to perfection. To allow for faster times and greater improvement, smaller divisions of time need to be created to monitor the race and the speeds of the runners. The addition of extra demarcations of seconds down to a tenth, to a thousandth, to a millionth of a second allows for more records to be broken. However, the quantifiable interval does not bring us closer to understanding the paradox of time. The addition of increments re-enacts Zeno's paradox, which determines an understanding of space as being divisible and positions us as the arbiters of that division (Huggett 2011).

In the example where an arrow fired from a bow has to reach a target, then each space that the arrow passes through is divisible, whereas an instant is indivisible. Thus space cannot be an instant. Bergson critiques Zeno's paradoxes in relation to indivisibility, he explains: 'movement is articulated inwardly. It is either an indivisible bound (which may occupy, nevertheless, a very long duration) or a series of indivisible bounds. Take the articulations of this movement into account, or give up speculating on its nature' (Bergson 1911: 327–328).

In this statement, Bergson sees the indivisibility of all movement from its own articulation. So Bergson contesting the paradoxical nature of measurement brings into question the scientific rationale for the deployment of quantifiable information. The idea that reality is not constructed of static or frozen moments masquerading as states of being, gives way to the challenge to our understanding of life and its unpredictable potential being more than a movie. The unpredictability of measurements is reflected in Marcel Duchamp's artwork *3 Standard Stoppages* (1913–14).[5] The work demonstrated a challenge to certainty, allowing the potential for creativity and intuition to play a role. We might also see a challenge back to Marey's chronophotographic experiments and their assimilation of movement from frozen moments on film:

> These preoccupations were clearly evident in the work of Marey who was concerned with the physiological foundations of movement rather than the appearance of things. Marey, like Janssen and a number of other important figures in the technological prehistory of the cinematograph, addressed the epistemic uncertainties that induced a distrust of the observer in scientific experiments. Since it was thought that the eye could so easily be confused, observation without instruments was distrusted. … in the pre-cinematic age when Etienne-Jules Marey, for example, inquiring into the nature of movement, regarded the new techniques of chronophotography as inferior to graphic methods using smoked drums and scribes attached to pneumatic sensors. Phototechnology used shutters that insisted upon the moment as an event of finite duration, an insistence that consequently ruptured the flow of movement as experienced in a flux of time.
>
> (Punt 2002: 61)

The instances of time that were measured by Marey show how the machine is used to interpret phenomena of its time that seems inexplicable to human comprehension. Marey's experiments are analogous to the division of the material world at a nano level where every atom is presented as an instance in time. Similarly our understanding of space is relative in this context as we consider matter at an atomic level as being empty. At the atomic level there is more space than substance.

During the turn of the century the atomic space was referred to as 'ether' and was thought of as a medium that filled all invisible space. Ether as a physical manifestation is incorporated and visualized in Boccioni's *Man forms unique continuity in space*, as mentioned in Chapter 2. The ether that Boccioni represented can now 100 years later be related to a new interpretation of energy that permeates 96 per cent of space and is referred to as dark energy and dark matter. The dark energy/matter is now being visualized in a similar context to that of Marey's absence of light. Light absorbed in the black background is not captured by the light sensitive film but made visible as a by-product of the movement that it surrounds in the photographs.

While all of Zenos's paradoxes of motion can be linked to a reductionist theory of a mathematical atomistic space, the desire to see all atoms in a state of motion, and demonstrate their complexity through measurement by technologies like the AFM, is brought into perspective between the positions taken by Zeno and Bergson.

## *A Mote it is ...*

In the work *A Mote it is ...* Mike Phillips tackles ocularcentrism that prioritizes the eye epistemologically, by taking a tiny speck or particle (Mote) from the artist's own eye as the sample substrate to be studied microscopically. The sample mote was scanned at the Wolfson Nanotechnology Laboratory at Plymouth University in collaboration with Professor Genhua Pan. The AFM (a Pacific Nanotechnology Nano-R) was set in contact mode to gather topographical coordinates.

The bringing of something into being is the main conceptual trigger of the work. The mote acts as interference in the corner of one's eye between what is seen and what is perceived. The work recognizes a 'mote' as both noun and verb, meaning both the smallest thing possible and to conjure something into being (*so mote it be ...*). Phillips suggests: 'This dual state of becoming and being (even if infinitesimally tiny) render it a powerful talisman in the context of nanotechnology' (Phillips 2010).

The ghost of Hamlet's father provides a metaphysical foil for the piece, *'A mote it is to trouble the mind's eye'* (Shakespeare). This highlights the impossibility of seeing something that is not there (maybe) and hints at the tension in our desire to see the invisible. The mote is both a filter and an irritant. This focus on the smallest of particles that acts like a filter at the periphery of the eye's surface generates an almost recognizable pattern. The mote creates patterns that hover between visibility and invisibility, being and not being. Phillips suggests: 'our view of the "real world" is increasingly understood through images made of data, things that are measured' (Phillips 2010).

A mote is the particle of dust that once caught in the sunlight becomes visible and the vortices of air create patterns and swirls. This brings us back to the unpredictability of Lucretius' swerve that is self-evident here. The motes float around in a state of chaos and from that chaos we have a potential to see forms appearing. The atoms at play in the work *A Mote it is ...* are created from the AFM's data, where an instance of each individual atom's coordinates can be recorded. The viewer can only see the evolving patterns in the installation appearing on the periphery of perception. Face recognition software, incorporated into the installation, recognizes the viewer's fixed stare as they advance to get a closer look. On being seen the mote degrades only to become invisible. The face recognition software provides an interface to the visualization. *A Mote it is ...* recognizes when it is being looked at and disappears, ghost like, resorting to silence and a raw data state.

*A Mote it is ...* ignores the images generated by the AFM, instead the raw data from the scan is used as a source for a particle cloud generated using a Flash Media

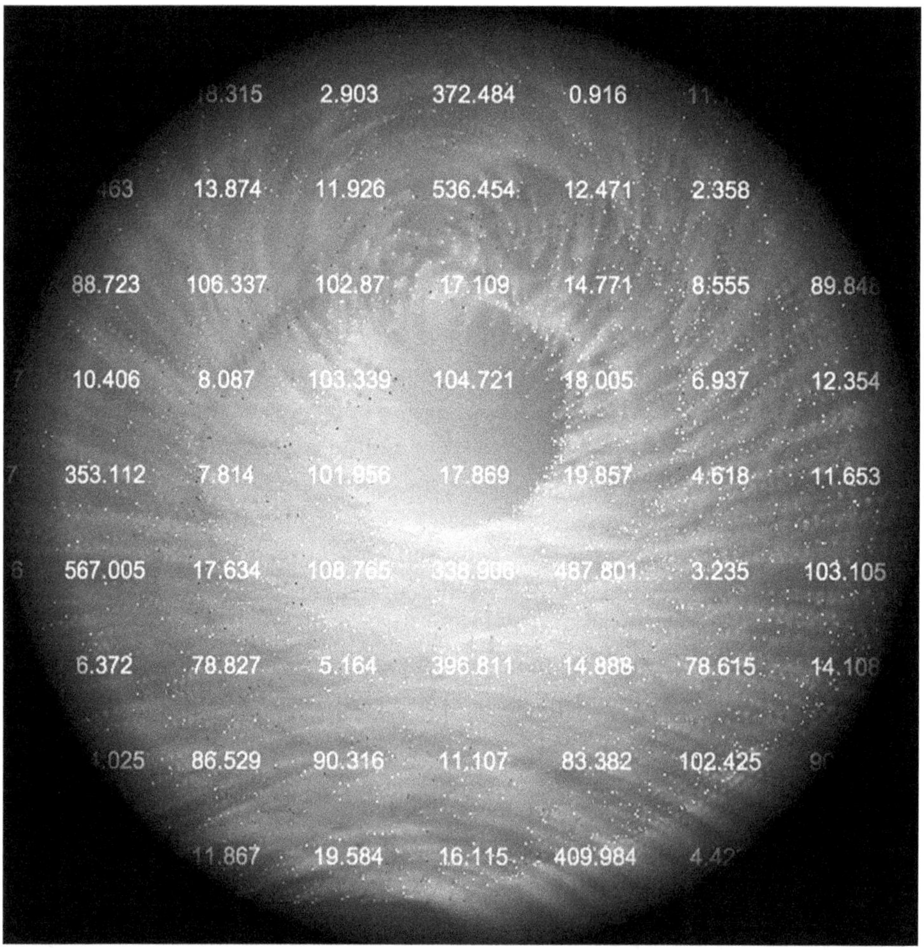


**Figure 1:** Mike Phillips, *AFM Mote* (2009). AFM digital image. Courtesy of the artist.

Server application. The data from the scan, which originally took 15 minutes for the AFM to acquire, is reconstituted in a whirling cloud of particles. The swirling cloud is accompanied by a sonification of the same data. The audio was created by generating a sign wave from the vertical topography of the scan whereby a nano landscape is transformed into a linear track. The audio was transposed through various frequencies to give a composite sound. The duration of the audio and particle cloud (uninterrupted) is the same as the original scan.

Like Hayles, who references pattern as the realization of a 'certain set of possibilities', Phillips takes the dust or skin particle floating in the ether and makes it the source for further investigation offering a new understanding of the invisible substrate of the world. He provides a new perceptual mode that privileges the periphery and where haptic interfaces allow us to see the invisible. The mote is the cause and the AFM is the mechanical interface that creates the recorded data (Figure 1). Philips' work enables the viewer to see data returned to a humanist discourse where recognition of the material world is once again interpretive.

## *Hostage*

The exploration of a scientific understanding of measurement is relative to the work of three artists working with elementary particles. For Frederik De Wilde, Mike Phillips, and Paul Thomas in collaboration with Kevin Raxworthy, elements like the photon are a means of expanding our awareness of the immaterial and material world made visible.

A fuller understanding of matter in relation to the questions raised by nanotechnologies requires a social adjustment or trigger to create a rupture in the fabric of human perception of light, time and space. The history of art from the mid-nineteenth century is strongly aligned with the pursuit of scientific discoveries. However, artists were never blind to what was being explored both in the sciences and in parallel developments and critiques in relation to social and cultural change. Similarly contemporary artists' critique of nanotechnology is in part through qualitative research, where the works of art transcend what has been made quantifiable by science.

The challenges taken up by artists dealing with scientific quantifiable research in the area of nanotechnology can be seen in an iteration of the artwork *Hostage* by Frederik De Wilde. The work builds on scientific research led by Shawn-Yu Lin, professor of physics at Rensselaer, Polytechnic Institute and Professor Pulickel M. Ajayan at Rice University in Texas, Houston, USA. *Hostage* explores the concept of measurement by examining the use of nanotubes and their connection with contemporary thinking

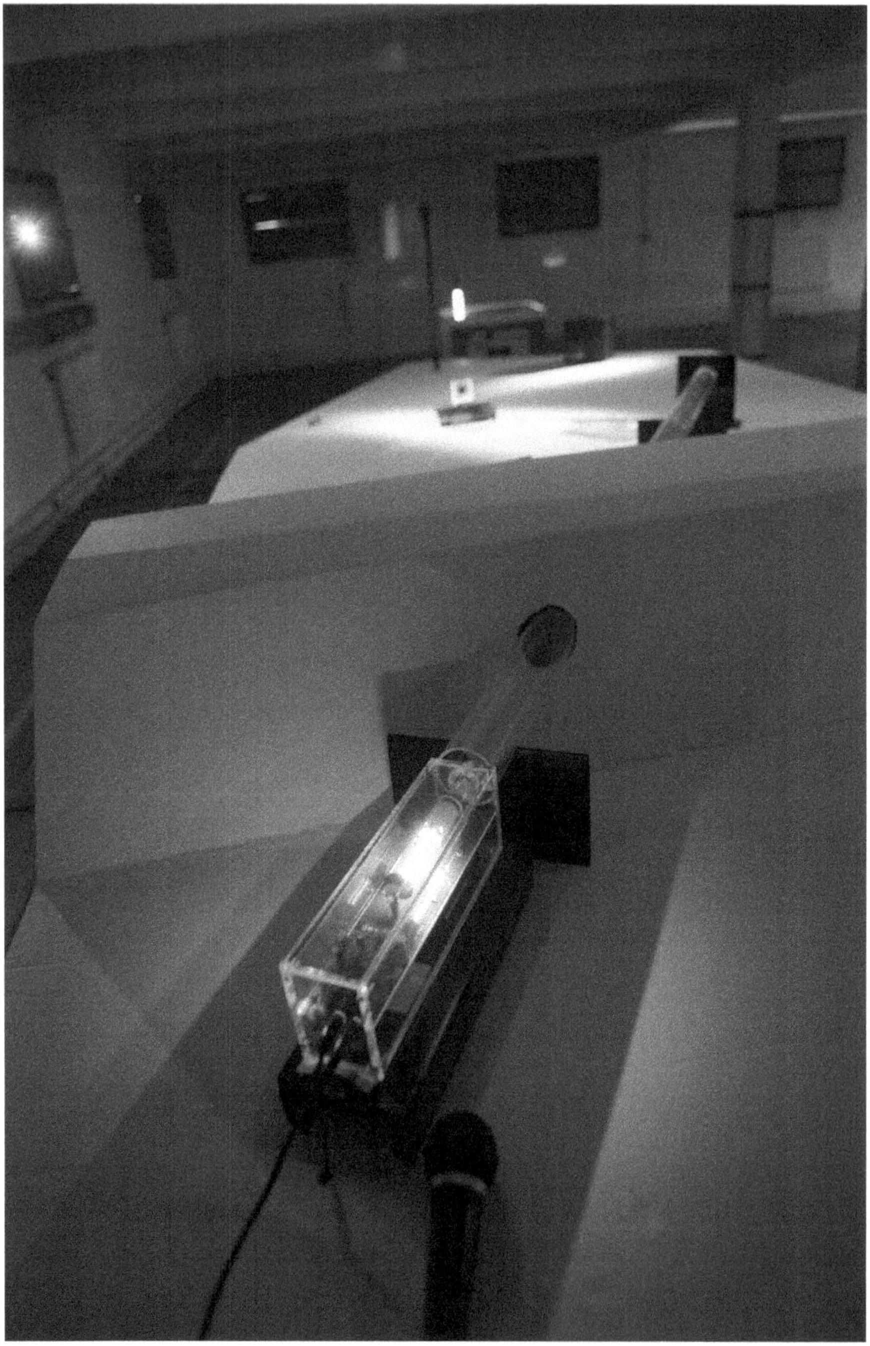

**Figure 2:** Frederik De Wilde, *Hostage* (2010).

about matter and light. *Hostage* explores the darkness, or blackness, created by an array of vertically aligned and loosely packed carbon nanotubes (CNTs).

The artwork is made with nano-sized metallic particles deposited on a substrate such as seed catalysts and grown into a 'forest' of CNTs by the means of chemistry. When light photons hit materials like paper, water or mirrors, a certain amount of light is absorbed and the rest is reflected whereas in the research lab CNTs are clustered together to form a mass, which is used to collect and trap all light. The concept of these clustered CNTs is to create a super black object that would absorb all light in not only the visible, but also in the invisible domains (infrared and far-infrared).

In Chapter 1, I discussed how nanotubes can be linked in an art historical sense to graphite and to drawing. The work is relative at an artistic level to the carbon that one finds in charcoal or in pencils (graphite). De Wilde uses blackness gained from the materiality and effect of nanotubes and theoretically embeds it into the history of art (Figure 3). He references fine art painters' continual battle with blackness and with the absence of light that is so essential to maintain a pictorial ocularcentric understanding of the world. The light allows for darkness in the form of shadows to give the representational image its clarity and depth. It plays on the metaphor of blacker than black and demonstrates De Wilde's contemporary proposition that colour creates a direct link to the ancient practice of drawing by the first human beings in the Caves of Lascaux, France, and to the beginning of the twenty first century.

The use of mythology in De Wilde's practice-led research allows for connections to be made with a primordial understanding of a situated knowledge of blackness. One of the starting points of De Wilde is the *Helmet of Darkness* that comes from Greek mythology. Haides (Hades) along with Posideion (Poseidon) and Zeus were given gifts by the Kyklopes (Cyclopes) to defeat the elder Gods in a divine 10-year battle called the Titanomachy. Haides (Hades) was given a gift of the helmet of darkness. The helmet created invisibility and enabled the user to move about undetected. The helmet later came to be associated with deception. When the helmet of darkness was worn, it had the potential to absorb all light photons that hit the surface of the individual rendering their mass invisible.

Black has been a concern for painters and De Wilde makes reference to a number of them. For example, Kasimir Malevich's *Black Square* (1913) is a pure abstraction and a seminal example of the loss of representation and the power of the object. In this work, all objects are absorbed into a black hole as an expression of a pure void and absence. The black of the oil paint has consumed all the light and left us with the invisibility.

The prototype work *Hostage* presented at Ars Electronica in 2010 sonically translated the concept of absorption. A vocal sound was transformed in to light in the form of a laser beam that was then aimed at the sample of nanotubes (Figure 2).

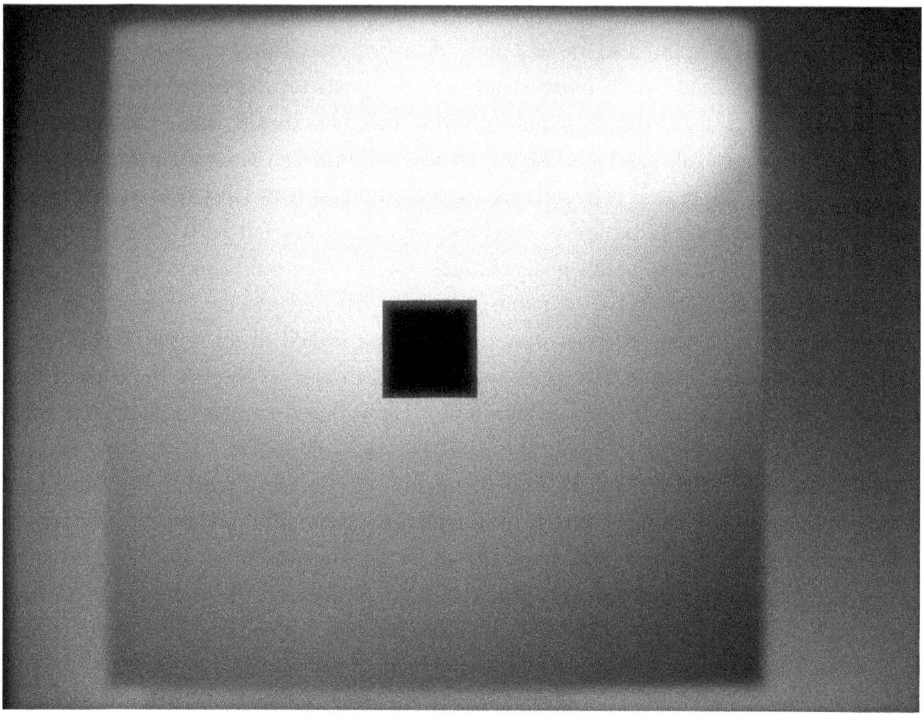

**Figure 3:** Frederik De Wilde, *Hostage* (2010).

The residual laser light deflected from the surface was picked up by a sensor and translated back to sound. Hence a sonic iteration could be perceived of the residual produced from the nanotubes' ability to absorb light. The use of laser light in *Hostage* mimics the AFM's use of the laser to record the subtle movement of the cantilever as it makes atomic contact with a sample surface. The cantilever's motion is recorded as the laser deflects onto a photodiode and its varying positions are translated into numerical text files. De Wilde's work draws attention to the problems of recording and visualizing atomic data due to the loss that occurs. The work uses technology to correlate the sense of touch with sound to entice the viewer into a dialogue with the material absorption of data at a nano level. *Hostage* examines a machinic understanding of the world through the exploration of the nanotubes' absorption of photons that create a maelstrom of blackness. Marshall McLuhan uses Edgar Allan Poe's *A Descent into the Maelström*[6] as an example exploring things within the storm that disappear and can also reappear; from the recognition of these patterns made by the reappearance, new ways of perceiving and understanding materiality can be fostered.

## *The Claude Glass*

In relation to the loss explored in *Hostage*, we might draw parallels with *the Claude glass* (Black mirror) a device that directly deals with loss and the mirror. The Claude glass was named after its inventor, the painter Claude Lorrain (1778) and was widely used in the seventeenth and eighteenth centuries by landscape artists. The small convex mirror had a darkened surface that created a filter between the landscape and the artist. The artists would turn their backs on the view only to be seduced by the muted tones in the mirror. Paintings were made directly from the fetishist screen of the darkened mirror. What was lost in the process was not only details of the landscape but also a personal response.

The photons reflecting back through the blackness represent the concepts explored by De Wilde and Thomas, where the loss of the subject is recorded and translated back into a perceivable form. Similarly, we might think of loss in relation to the switched off computer screen, like the screen of the television, it has the same power to display through reflection, a seductive and unattainable representation of the world.

Before discussing another work that relates directly to the loss in relation to light and use of the mirror, it is useful to introduce a work that problematizes the notion of seeing in relation to nanotechnology.

## *Atomism*

Thomas and Raxworthy, who are exploring what is substituted for the loss that occurs when light is absorbed and reflected from a mirrored surface, can link De Wilde's work dealing with loss with the work of *Atomism*. *Atomism*, like *Hostage*, is situated in an art historical and theoretical model of enquiry into nanotechnology. *Atomism* is an installation that investigates silver, the mirror and quantum theories of light. Feynman, in his 1979 lectures, tells us that light hits a mirror at all points not just at the point of reflection (Trust 1979). All points on the surface of the mirror receive and reflect light based on the speculative spin of the photon that is not visible to the viewer. The premise of the work reveals that a light wave does not merely encode all the information about an object but is also independent of an observer. The diagram showing the analogous spin is based on Feynman's 'probability amplitude'[7] that demonstrates the photons as only having one visible position. This visible position is the shortest angle of incidence, where the reflection is made visible to the viewer. *Atomism* reveals that all the possible alternative 'quantum worlds' have the potential to be equally real and exist simultaneously (Figure 4).

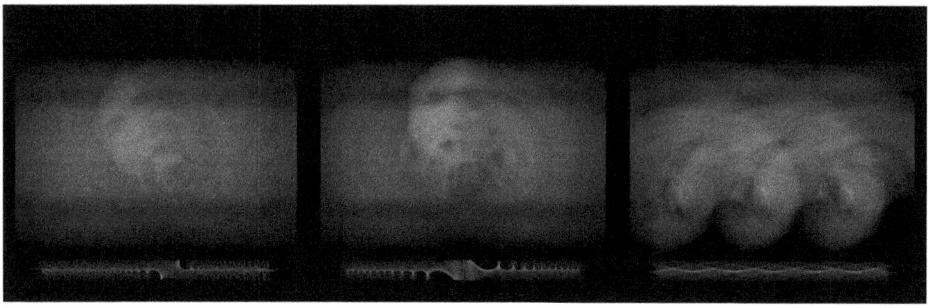

**Figure 4:** Test image for *Atomism* (2011). By Paul Thomas and Kevin Raxworthy.

The concept for *Atomism* uses a data image from the AFM of a 7-nanometre scan of a silver substrate that acts as a filter for the live camera feed image of the viewer's self-portrait. Each viewer's real-time reflected portrait is analogous to single photons. The photon's spin is mapped to reflect back to the viewer, showing all of the representations, not just that closest to the angle of incidence. The computer generates nearly two thousand versions of the viewer's image stacked, which is the equivalent to Feynman 'summing' the angles of probability. The mathematics is processed to react to the 'probability amplitude' of the photons' paths across the screen of the metaphorical mirror. The diagram shows the spin of the photon with the shortest axis being at the centre point. As the spin is visualized along the entire surface, there are points of clarity where the multitude of reflected spinning images coalesce. The mathematics in visualizing the self-portrait reveals a repetitive simulation of the occurring original image. These repetitive simulations of images created in the work *Atomism* are likened to that of Schrödinger's cat referenced in Chapter 1; they come into existence in more than one place at the same time. The experimental research for the work discovered the potential for parallel universes, multiverse or many worlds[8] that sit at the extremities of the photons' spin as they are reflected from the surface of the metaphorical mirror.

The work creates a visual data cloud from the spins whereby emerging forms that are made visible along the spectrum can be glimpsed. The viewers see themselves and glimpse within the pattern and randomness, a potential for the emergence of forms thereby creating a series of becomings. The perception of these patterns and randomness that form within the space of the parallel universe of a mirrored surface can be considered as the basis for new knowledge to be revealed. Hayles relates to 'seeing randomness not simply as the lack of pattern but as the creative ground from which pattern can emerge' (Hayles 1999: 286). The series of becomings within the mirrored surface are potentially multiple liminal spaces between the many possible patterns created from our understanding of quantum physics phenomena.

There is the potential for more patterns to be recognized that are born out of the unpredictability of randomness and chaos. Therefore the recognition of pattern from the unpredictable allows for the world to be seen as more than it appears. Insignificant realizations and phenomena can be explored in the work with greater freedom for the potential of something coming into existence. The recognition of patterns embodying presence is not understood as an isolated act but is linked to a total ecological network of experiences.

Historically, there is the recognition that an object's presence in space is created out of a form of pattern recognition, this can be seen in contrast to the disembodied nature of a computer image, which is made out of a pixelated, constructed pattern.

Both of these visual experiences of presences and pattern are seen in a context of a dominant perspectival understanding of space. The perspectival point of view is an objectification of space where the potential for the recognition of presence and pattern from randomness is thwarted. The point of view separates the viewer from what is seen or felt. The data that creates the computer vision in its binary state is viewed as randomness waiting to evolve into pattern. Hayles points out when linking pattern to presence in the non-material space of the computer that 'cyberspace defines a regime of representation within which pattern is the essential reality, presence an optical illusion' (Hayles 1999: 36). Current scientific images all constructed through a digital machinic process are based on pattern recognition. The images like all others taken through a digital device are pattern being seen as presence.

These contextual and conceptual processes are manifested in the work *Atomism* by initially scanning burnished silver to explore its atomic pattern and to discover if anything is lost in the process of reflection at a nano level. Dr Thomas Becker from the Nanochemistry Research Institute, Curtin University, using the STM and AFM, gathered data of silver samples as a basis for the pattern/randomness visualization. The silver was burnished and scanned, then heated to give the sample a smooth surface needed for detailed scanning with an AFM.

The initial interest in the use of silver was based on its properties to act as a mirror. Silver played an important role in the birth of perspective and the objectification of the world from a renaissance perspective. Fillipo Brunelleschi used polished silver in his 1425 peephole device to replace the sky in his painted image of the baptistery outside the Duomo in Florence. As Brunelleschi's biographer Antonio di Tuccio Manetti points out, after having witnessed the device, 'the sky had to be represented, … the buildings of the painting were free thus the clouds seen in the silver are carried along by the wind as it blows' (Manetti 1970). The need for perspective to deal with nature was not necessary for the architectural requirements of Brunelleschi. Therefore, the residual image of the sky reflected in the silver of Brunelleschi's painting became, by default, perspectivalized resigning nature to be objectified and dominated by the individual gaze. The silver used in the painting

> is also a mirror, in this case a mirror that captures what perspective cannot. Thus even what is in excess of perspective's vanishing point is brought into its hegemonic screen. Retrospectively, we might today see in the burnished silver the harbinger of virtual reality and of photography; though, of course, this is exactly Lacan's point about the psychology of the mirror and, by inference, of perspective itself. Brunelleschi's demonstration established a new spatiality, which inaugurated the hegemony of the virtual over the actual. Perspectival space was the new real.
>
> (Thomas 2009b)

The perspectival space was made real partly through the reflective qualities of silver mirroring nature to the viewer. Michel Foucault's (1986) 'heterotopia'[9] is evidenced here in the viewer who sees the clouds moving by and recognizes they have been processed as reflections from somewhere else. The pattern of the cloud presented through the device is recognized as being real but also virtual. The creation of Brunelleschi's illusion in the peephole device is part of a genealogy of the first camera and various viewing devices.

Another reason for the choice of silver in the work relates to its material quality. Silver, when dissolved in nitric acid and processed, turns black and grey when exposed to light. The relationship of silver to pattern and randomness in the history of art is evident through photography and the emergence of an image from the reactions of photons hitting each grain in the film stock.

The Pointillist artist George-Pierre Seurat, who was painting and drawing around 1884, focussed on the grain as the basis for the exposure of a representational image. He is significant historically in exploring work that is de-territorializing the world. The idea of seeing Seurat as a touchstone for Nanoart is made by situating his paintings and preparatory drawings as being related to concepts of becoming and presence. The representational images are produced through an array of coloured dots of paint and graphite textures that form patterns that coalesce in the eye. Seurat's preparatory drawings and his paintings point to a time where matter was unstable and viewed as part of the ether, highlighting 'the atomic movements at the deep structure of nature' (Milburn 2010).

Seurat's paintings construct the window through which the distance between us and the world of atoms is calibrated to give the world presence. What can be seen in Seurat's work is a resonance with the 'clinamen', where the independent swerve of atoms operates in a void that takes place beyond our perception. In Seurat's painting, *A Sunday on La Grande Jatte* (1884–1886), the forms emerge from the chaos presented by a border that surrounds the image. The border takes the primordial chaos of the world and shows it evolving into recognizable forms. Seurat's political aspirations for the painting *Bathers at Asnières 1884,* seems to reinforce the currency of giving the world presence and pattern by recalibrating the de-territorialized boundaries that separate individuals from the material world.

Seurat had an interest in science as indicated by many of his experiments in colour from the findings of French chemist Michel Eugène Chevreul. Chevreul was instrumental in identifying the effects of colours' proximities to one another in his work on the restoration of tapestries. He noticed that the colours in the tightly gridded space of the tapestry affected each other's value and colours needed to be assessed in relation to their surroundings to determine the correct colour value.

The rethinking of Seurat's work, which was derived from scientific exploration of tapestries, is a precursor of the digital age where each pixel is analogous to the atom. Therefore the focus on Seurat's intuitive work from a nanotechnological perspective gives a different contextual agency for the works. These ideas can also be relevant in situating the work of Impressionist artists such as Claude Monet, where if we look at impressionistic work not from the point of view of the 'fleeting moment', but rather as an intuitive understanding of light's de-materialization of matter, scientific research is being explored and presented. The photons hitting Monet's Rouen Cathedral paintings (1892–1893) literally de-territorialize the building, turning matter into soft melting forms. Light becomes matter and matter becomes light. These works are, amongst others, prime examples that pre-date or are synchronistic with a rethinking of time and space. This de-territorialization would later be related to a Drexlerian concept of matter being visually de-territorialized by photons.[10]

## The photon

The photon's frequency and spin challenges the perception of materiality in paintings that appear to be dissolving as a result of the photons' absorption into and reflection off the surface. This seemingly fluid effect of repositioning painting and sculpture in the context of theories of light and matter demonstrates that artists were not only bringing an element of seeing the world into focus, but also showing an intuitive and innate understanding of its causal molecular affect.

During his 1979 lecture Feynman presents information to the audience when describing an arrow used in calculating the 'probability amplitude' as being

> like a very rapidly moving rotating clock hand going along as long as you wish, for the length of time that you would calculate ordinarily for the light to come to the first surface and bounce back. It goes around like a sun of a gun .... one followed by 15 zeros per second.
>
> (Trust 1979)

Feynman's diagram demonstrates an arrow's spin as being analogous to the probability of the spin of a photon travelling at the speed of light while rotating at a rate of fifteen zeros, which equates to 40 trillion revolutions per second. I have previously written about how the spin is used in the processing of information in quantum computing. The photon with its spin based from the 'probability amplitude', which works at a speed beyond human comprehension, records information that when absorbed into the material world could be constantly archiving the visual world. The photon's energy is immersed into the materiality as a steady stream of waves or particles.

Therefore the potential exists to explore what happens if the spin of the photon could collect data, and if the data could be absorbed into matter. The information recorded by the photon of the object on its travels then transposes the energy absorbed as it collides into a material. The potential is to look at how much energy is absorbed and whether information can be reverse engineered.

## Matter

Similar to *Hostage*, the works *A Mote it is* ... and *Atomism* are reflective and evolutionary as data is processed and remediated creating intimate nano experiences for the viewers.

The art historical contexts for these installations demonstrate a relationship with a syncretic understanding of shifts in material agency brought about by nanotechnological research. These artists through their practice portray a sense of a bigger meaning in the 'nature of things' by recontextualizing our new concerns, and reconfiguring our fundamental relationship with matter.

The recontextualizing and repositioning of the human away from a dominant ocularcentric view of the material world points towards Lucretius and the unpredictable. What artists have shown in these chapters is an ongoing commitment to explore the instability of perception leading to a rethinking of the quantum theory and the potentiality of matter. The challenge in the fine arts has been to resist seeing the world from the perspectival point of view, to overturn the dominant gaze that reduces all matter to its servant. For example, in the work of Boccioni we have seen the human and the material object world coalescing, forming no hierarchical bonds. These bonds were being formed simultaneously with scientific rediscovery of ether as being a medium that permeates all space.

The desire for a shift to occur in the human to recontextualize our understanding of the agency and vibrancy of matter is the basis of the art works presented. The artists presented look at various interfacial methods of contributing to a discourse that has been part of a contemporary art tradition since the middle of the nineteenth century. Matter needs to be understood in a new light as one that is in a process of becoming. The same becoming that a Darwinist would envisage from micro-organisms developing in the fluid medium of water. The addition of various unpredictable elements at a molecular level creates chemical cocktails for the potential of matter to become. The chemical compounds that go in to making a physical body and therefore have the potential for life have a 'total street value of not much' (Amato 2000). This value is part of the problem of the readjustment of our understanding of matter. Not only is it hard for humans to deal with measurements beyond their immediate surroundings, but also to see the potential for more than what just appears on a material's surface.

## Notes

1 Deleuze writes in his book *Bergsonism* about Bergson's concept for nonbeing when he states that 'there is not less, but more in the idea of nonbeing than that of being, in disorder than in order, in the possible than in the real'. Deleuze, G. (1991), *Bergsonism*. New York: Zone Books.

2 Serres reinforces the primary parmenidean understanding of atoms by stating 'if it is absurd that a small solid mass might at some moment deviate from the orbit of its fall, let us examine whether the same may be said where the primary atomic cataract is like a stream, like a flux, like a flow of a liquid. Lucretius says elsewhere that the subjects of physics are mass, fluids and heat. And since for him everything flows, nothing is truly of an invincible solidity, except for atoms.'

3 A phase transition can refer to breaking symmetries in the laws of physics of the bonds that hold matter together.

4 Zeno was part of the Eleatic school with Parmenides who created paradoxes to question plurality and any kind of change, for him this was reduced down to one thing that could be defined by reason and thought.

5 Duchamp's *3 Standard Stoppages* is based on a piece of string a metre long being dropped from a metre high allowing the twists of the string to fall at random. The string was then glued to a canvas and cut out to form a template of a new metre in length.

6 The maelstrom is significant in pattern recognition and this reference is transcribed from Marshall McLuhan's last public lecture where he discusses Edgar Allan Poe's *A Descent into the Maelström* (Poe 1841) 'in the descent into the maelstrom Poe imagines the situation in which a sailor who has gone out on a fishing expedition, is caught by not watching the turn of the tide in a huge maelstrom or whirlpool. He sees that his boat will be sucked down into this thing; he begins to study the action of the ström, and observes that some things disappear and some things reappear. By studying those things that reappear and attaching himself to one of them, he saves himself. Pattern recognition in the midst of a huge, over-whelming, destructive force is the way out of the maelstrom'. The Marshall McLuhan speaking extract has been taken from a live audio recorded lecture by Derrick de Kerckhove, *McLuhan still dead? 25 years later*. http://www.aec.at/en/festival2005/podcasts/podcasts.asp.

7 Feynman in his lecture explains the probability amplitude in relationship to the spin of the photon. 'The probability of any event in an ideal experiment – that is just an experiment in which everything is specified as well as it can be – is the square of something, which in this case I have called 'a', the probability amplitude. When an event can occur in several alternative ways, the probability amplitude, this 'a' number, is the sum of the 'a's for each of the various alternatives. If an experiment is performed which is capable of determining which alternative is taken, the probability of the event is changed; it is then the sum of the probabilities for each alternative.' Feynman, Richard (1994), *The Character of Physical Law*, Modern Library.

8 The theories of many-worlds, parallel universes and multiverse are likened to 'Schrödinger's cat' thought experiment that implies the possibilities of alternative realities each representing an actual world. The theory was originally formulated by Hugh Everett in 1957.

9 The mirror functions as a heterotopia in this respect: it makes this place that I occupy at the moment when I look at myself in the glass at once absolutely real, connected with all the space that surrounds it, and absolutely unreal, since in order to be perceived it has to pass through this virtual point that is over there. Foucault, M. and Miskowiec, J. (1986), 'Of other spaces', *Diacritics*, 16: 1, pp. 22–27.

10 In his book *Engines of Creation* Eric Drexler forecasts a 'gray goo' that can de-territorialize matter. 'The gray goo threat makes one thing perfectly clear: we cannot afford certain kinds of accidents with replicating assemblers'. Drexler, E. (1986), *Engines of Creation: The Coming Era of Nanotechnology*, New York: Anchor Books.

# Chapter 5

## Transvitalism and Nature

> The universe of Epicurus and Lucretius is a reconciled one in which the science of things and the science of man go hand in hand, in identity. I am a disturbance, a vortex in turbulent nature. I am an ataraxia in a universe in which the heart of being is undisturbed. The wrinkles on my brow are the same as the ripples on the water. And my appeasement is universal.
>
> (Serres 1982: 121)

What I have attempted to do in the previous chapters is position the artworks presented as indicators of intuitive, synchronistic and serendipitous artistic insights that parallel the world of science. The artworks presented here are assembled according to their attempts to reconfigure human perception, connectivity and comprehension of matter. Bennett points out that a sustained relationship with 'the vitality of matter as [is] real' is difficult, she explains: 'it will be hard to discern it, and, once discerned, hard to keep focused on' (Bennett 2010). The artists presented here are compelled to not only discern and synthesize, but to translate and visualize in real time a view of the material world made quantifiable by scientific discoveries of the molecular. The sense of what constitutes matter and its expectancy for 'being' in the world demands through meditation our inclusivity and interaction with it. Not as 'other' but as a material coalescence with the materiality of being human.

The materiality of being human is contextually relevant to Deleuze and Guattarri's philosophical writings on the *refrain*.[1] The refrain highlights the link between metaphysics and science, creating a platform for interpreting the vibration of atoms to be at the core of materiality. The atomic milieu is brought about by vibration and pattern, paralleling Crichton's post-biological state of the human body where each organ is conceived as a swarm of atoms. Each atom that falls and then takes on the unpredictability of Lucretius' swerve connects with another atom to create a milieu. The milieu connects to other milieus that are formed out of the swerve and held together through their own forces of frequency and vibration. Serres when exploring unpredictability reflects 'that the turbulence deviates from equilibrium. And the beginning of the vortex is the minimal angle of declination. The fact that life disturbs the order of the world means literally that at first life is turbulence' (Serres 1982:102). The discovery of the perturbation in the Higgs boson mentioned previously can be added to Serres perception of a deviation, a turbulence being the first life.

The concept of 'assemblage'[2] is where each object, body, thing is represented as molecules swimming in the milieus. Each milieu forms a synergy with another milieu affecting its state, its multiplicity and making changes. The 'assemblage' is most relevant to discuss at this point in time because it is 'itself the medium' in which we think and act, in which we swim. We need a concept that is amphibian, and that can leave its medium (Thomas and Colless 2010). The assemblage as an artistic medium engulfs the ubiquitous images, transforming them to act like semi-autonomous nano particles that become integral to the swarm: the 'mycora'.[3] The swarms become foundational for the assemblage to form new connections and patterns that will be recognized to be brought into our consciousness.

The view for a future understanding of matter is to accept the turbulence, which is the swerve, 'where life disappears from the closed interior of organisms and dissolves into molecular informational processes, code, feedback systems and programs' (Milburn 2005: 287). This comprehension of matter is what Richard Doyle calls a 'postvital' point of view, where life moves into the information systems denying the 'autonomous interiority of an organism' (Doyle 2003: 20.1). This view of matter as having a postvital future that is tied up with the quantifiable predictability of code as the medium creates a revision of control and order. The honouring of the swerve of Lucretius needs to resurface to confront this point of view. Life and matter are not just a process of a bio-engineer's logic whereby they create their own rationale for its creation and production. If matter has the potential for a life force then that alone should be sufficient for us to review our discursive relationship with it. We need to move beyond the ocularcentrism that projects an ethnocentric image onto matter totally based on a predetermined cultural understanding of materiality.

The 'postvital' future is one that is based on quantifiable data, not the unpredictability of the swerve. I would propose a transvital approach where all matter-energy, code, viruses, air and water are seen as relevant and related. What has been given to us by nanotechnology data is a demonstration of material ownership under a Cartesian and Newtonian[4] logic. We might reflect upon the first paragraph of the report into nanotechnology for the US government which stated that the entire amount of material that constituted the human body had a 'total street value: not much' (Amato 2000). This might be a factual account of the bodies' chemical compound, but it does not demonstrate that the matter itself is part of the unpredictability in the world of atomism. The context for the report seems very clear, it is to have control over all material things and give the government, through science, the authority to manipulate and control all matter. There is a need to explore the ecological possibilities that would create a conscious change in our attitude towards matter. After all as Serres suggests:

> as we misunderstood Lucretius because we were the children of Plato and the Stoics, because the fundamental facts of Epicurean nature remained marginal in traditional science, which was really not very Archimedean. From that point on, we ruled them out of the game in the history of science. Moreover, we put their nature outside nature, placing them in the soul and the subject. On the contrary, however, these facts are the foundation of materialism.
>
> (Serres 1982: 103)

Not only are these the facts of materialism, but also the facts of our understanding of matter within the world. What these chapters have focussed on is the way that artists interpret a need to redefine our understanding of the world. The critical vitalism of Bergson sees agency and life in all matter as part of a solution. This has influenced a generation of artists to pursue new concepts of material engagement. The rationale for shifting our understanding of material agency is to see it as our most significant relationship and imperative in the sustainability of the worlds' matter. Matter at the atomic level is the medium that we all swim in in the ocean of our being. This view is not to be a substitution for science, but to recognize the potentiality of the swerve and its significance in disrupting the mechanization of humanity. That is, to allow unpredictability to have a place alongside what appears to be quantifiable facts, to break down human anthropomorphizing in the context of matter. Non-human material elements are of this world and have the right to have their place in the evolutionary chain.

The hegemonic human position in perceiving the world as something that needs to be changed, developed and controlled is beginning to unravel. The problem is we cannot stand far enough back from matter to be able to make a critical stance towards it, we need to find a method of mediation. If we cannot stand far enough away from the spectacle, we cannot make any changes as there is no escape. In other words, we see from a perspectival point of view a sector and mistake this sector for the whole problem, 'we merely superimpose the character of wholeness on to the sector, the result being the familiar "totality"' (Gebser 1985: 18). A different holistic paradigm must be transmedial, transcoded and transcendental to encapsulate being in the world.

The material worlds visualized by the artists referenced in this book are a testament to a new vanguard of thinking about matter at the atomic level. The work encourages us to experience a shift that has been evolving through the theories and practices of these artists. Relocating artists' parallel serendipitous understandings of the scientific exploration within the art gallery critically engages us in a humanities discourse to reveal alternative understandings of the way art affects and creates new ways to look at life.

The discourse is to question our reliance on any one particular process in understanding the world around us. Taking a mechanistic Newtonian understanding of the world and relating it to Bergson's approach of seeing all matter as being in a playground of probability. The Feynman 'probability amplitude' is used to function in areas where things cannot be explained, the spin of the photons and what is made visible and what is invisible. The phenomena of the swerve is played out over again in what can be seen and understood of our being in and experiencing the world. To believe that humans are inextricably and profoundly linked to all forms of being and becoming and the life that is linked to being is one that cannot be measured or quantified. The problem of life we fail to address is whether life itself can decide about life as it is being lived. It is the same argument of the individualism that according to Simondon, predates the individual. Measurements evaporate in the context of a new life as the criteria are stripped away by the unpredictability of being, and the ineluctable changing states of our comprehension of matter.

## Notes

1 The refrain brings together a series of vibrational elements – sights, sounds, rhythms, material objects, geographical features, its own bodily reactions – into a provisional totality, a territory that now locates expressive qualities – colours, textures, tones, tempos – compacted into a kind of assemblage, a natural creation. Grosz, E. (2006), 'Vibration. Darwin, Deleuze and the Music for the Cosmos', *Constellations*, Sydney.

2 The context for the assemblage here places all the various contributing factors of object and subject together to form a basis for new connections to be made. 'An assemblage is precisely this increase in the dimensions of a multiplicity that necessarily changes in nature as it expands its connections.' Deleuze, G. and Guattari F. (1987), *A Thousand Plateaus. Capitalism and Schizophrenia*, Minneapolis: University of Minnesota Press.

3 From the novel 'Bloom' set in the year 2106, voracious self-replicating nanomachines called 'Mycora' have consumed Earth and other planets of the inner solar system. The 'Mycora' becomes an assemblage of all the organic and nonorganic that works as a total entity, where the individuals have been collectivized.

4 When Newton invented his laws of mechanics many people took this to be the death of the free will concept. According to Newton's theory, the universe is like a giant clockwork, unwinding along a rigid, predetermined pathway towards an unalterable final state. The course of every atom is presumed to be legislated and decided in advance, laid down since the beginning of time. Human beings were seen as nothing but component machines caught up irresistibly in this colossal cosmic mechanism. Davies, P. (1983), *God and the New Physics*, Harmondsworth: Penguin Books.

# References

Alexjander, S. (2007), 'Microcosmic music – a new level of intensity', http://www.oursounduniverse.com/articles/microcosmic.html. Retrieved January 2007.

Amato, I. (1999), *Nanotechnology: Shaping The World Atom by Atom*, Washington, DC, National Science and Technology Council and The Interagency Working Group on Nanoscience Engineering and Technology.

Apollonio, U. (ed.) (1973), *Futurist Manifestos*, London: Thames & Hudson.

Ascott, R. (2004), 'Info-light: art, technoetics and the biophotonic network', *Ciberart: Challenges for a Ubiquitous Identity*, Bilbao.

Asher, R. (2002), 'Brain machine interface', *Converging Technologies for Improving Human Performance*, W. S. B. Mihail C. Roco, Dordrecht, the Netherlands.

Benatar, R. H. a. L. (1998), *A Practical Guide to Scanning Probe Microscopy*, DIANE Publishing Company.

Benjamin, S. C., Briggs, A. D., Britz, D. A., Gunlycke, D., Jefferson, J., Jones, M. A. G., Leigh, D. F., Lovett, B. W., Khlobystov, A. N., Lyon, S. A., Morton, J. J. L., Porfyrakis, K., Sambrook, M. R. and Tyryshkin, A. M. (2006), 'Towards a fullerene-based quantum computer', *Journal of Physics: Condensed Matter*, 18: 21, pp. S867–S883.

Benjamin, W. (1969), *Illuminations*, New York: Schocken Books.

Bennett, J. (2010), *Vibrant Matter: A Polical Ecology of Things*, Durham: Duke University Press.

Bergson, H. (1911), *Creative Evolution*, New York: H. Holt and Company.

——— (1996), *Matter and Memory*, New York: Zone Books.

Bishop, R. (2012), 'Research in the School of Chemistry', U. o. N. S. Wales, Sydney: UNSW: 7.

Boccioni, U. (1973a), 'Plastic dynamism', *Futurist Manifestos*, U. Apollonio, London: Thames & Hudson, pp. 92–95.

——— (1973b), 'The Plastic Foundations of Futurist Sculpture and Painting 1913', *Futurist Manifestos*, U. Apollonio, London: Thames & Hudson, pp. 88–90.

——— (1973c), 'Technical manifesto of Futurist sculpture', *Futurist Manifestos*, U. Apollonio, London: Thames & Hudson, pp. 51–65.

Boccioni, U., Carrà, C., Russolo, L., Balla, G., Severini, G. (1973d), 'Futurist painting: Technical manifesto', U. Apollonio, London: Thames & Hudson, pp 27–31.

Boukamp, P., Petrussevska, R. T. et al. (1988), 'Normal keratinization in a spontaneously immortalized aneuploid human keratinocyte cell line', *Journal of Cell Biology*, 106: 3, pp. 761–771.

Braun, M. (1992), *Picturing Time. The Work of Etienne-Jules Marey (1830–1904)*, Chicago & London: The University of Chicago Press.

Copeland, E., Bowley, R. et al. (2011), 'Sixty symbols', http://topdocumentaryfilms.com/what-is-the-higgs-boson/.

Cresswell, T. (2006), *On the Move*, London: Routledge.

Crichton, M. (2002), *Prey*, London: HarperCollins Publishers.

Deleuze, G. (1968), *Difference and Repitition*, New York: Columbia University Press.

Deleuze, G. and Guattari, F. (1987), *A Thousand Plateaus. Capitalism and Schizophrenia*, Minneapolis: University of Minnesota Press.

Doyle, R. (2003), *Wetwares: Experiments in Postvital Living*, Minneapolis: University of Minnesota Press.

Drexler, E. (1986), *Engines of Creation: The Coming Era of Nanotechnology*, New York: Anchor Books.

Drexler, K. E. (2007), *Engines of Creation 2.0: The Coming Era of Nanotechnology – Updated and Expanded*, WOWIO Books: 646.

ETC Group (2003), 'Green goo: nanotechnology comes alive!' *Communiqué*, no 77, (January/February).

Evans, R. (2011), 'Study rejects "faster than light" particle finding', *Reuters*.

Feynman, R. (1960), Talk at the American Physics Society's annual meeting, 29 December in *Engineering and Science*, California Institute of Technology, 23: 5, February 1960, pp. 22–36. Original magazine layout available at http://calteches.library.caltech.edu/47/3/ES.23.5.1960.0.pdf. Retrieved 15 June 2012.

——— (1979), *Fits of Reflection and Transmission – Quantum Behaviour The Douglas Robb Memorial Lectures*, Bristol: The Vega Science Trust.

Foucault, M. and Miskowiec, J. (1986), 'Of other spaces', *Diacritics*, 16: 1, pp. 22–27.

Gebser, J. (1985), *The Ever-Present Origin*, Athens, Ohio: Ohio University Press.

Geim, A. K. and Kim P. (2008), 'Carbon wonderland', *Scientific American*, 298: 4, pp. 90–97.

Gimzewski, J. K. and Vesna, V. (2003), 'The Nanoneme Syndrome: Blurring the fact and fiction in the construction of a new science', *Technoetic Arts*, 1: 1, pp. 7–24.

Goldblatt, M. (2002), 'DARPA's programs in enhancing human performance', *Converging Technologies for Improving Human Performance*, W. S. B. Mihail C. Roco, Dortrecht, the Netherlands: Klumer.

Grosz, E. (2006), 'Vibration. Darwin, Deleuze and the Music of the Cosmos', *Constellations*, Sydney.

Hansen, M. B. N. (2006), *New Philosophy for New Media*, Cambridge, MA: The MIT Press.

Hayles, N. K. (1999), *How We Became Posthuman: Virtual Bodies in Cybernetics, Literature, and Informatics*, Chicago: The University of Chicago Press.

Henderson, L. D. (1983), *The Fourth Dimension and Non-Euclidean Geometry in Modern Art*, Princeton: Princeton University Press.

——— (2004), 'Editor's introduction: I. Writing modern art and science – an overview; II. Cubism, futurism, and ether physics in the early twentieth century', *Science in Context*, 17: 04, pp. 423–466.

Hoffmeyer, J. (1994), 'The swarming body', *Semiotics Around the World.* Proceedings of the Fifth Congress of the International Association for Semiotic Studies, Berkeley.

Hooke, R. (1665), *Micrographia: Or Some Physiological Descriptions of Minute Bodies Made by Magnifying Glasses With Observations and Inquiries Thereupon*, London: Royal Society.

Huggett, N. (2011), 'Zeno's paradoxes', *The Stanford Encyclopedia of Philosophy* Winter 2010 Edition, http://plato.stanford.edu/archives/win2010/entries/paradox-zeno/%3E.

Kant, I. (1987), *Critique of Judgment/Immanuel Kant*, translated, with an introduction by Werner S. Pluhar; with a foreword by Mary Gregor, Indianapolis: Hackett Pub. Co.

Klee, P. (1961), *Paul Klee: The Thinking Eye: The Notebooks of Paul Klee*, London: Lund Humphries.

Lucretius Carus, T. ([c. 50 BCE] 1950), *Of the Nature of Things*, New York: Dutton.

Manetti, A. d. T. (1970), *The Life of Brunelleschi*, University Park: The Pennsylvania State University Press.

Marey, E. J. (1895), *Movement*, New York: D. Appleton and Company.

Marijuán, P. C. (2002), 'Bionformation: untangling the networks of life', *Biosystems*, 64: 1–3, pp. 111–118.

McCarthy, W. (1998), *Bloom*, London: Millennium.

Merleau-Ponty, M. (1968), *The Visible and the Invisible*, Evanston: Northwestern University Press.

Mignonneau, L. and Sommerer, C. (2005), 'Nano-Scape: experiencing aspects of nanotechnology through a magnetic force-feedback interface', ACM SIGCHI International Conference on Advances in Computer Entertainment, Spain, pp. 200–203.

Mignonneau, L. and Sommerer, C. (2010), *Nano-Scape Art in the Age of Nanotechnology*, Perth: John Curtin Gallery.

Milburn, C. (2004), 'Nanotechnology in the age of posthuman engineering: science fiction as science', *Nanoculture: Implcations of the New Technoscience*, N. K. Hayles, Bristol: Intellect.

——— (2005), 'Nano/splatter: disintegrating the postbiological body', New Literary History, 36: 2, pp. 283.

——— (2010), 'Tactical atomism', Art in the Age of Nanotechnology, Perth: John Curtin Gallery.

Miller, A. I. (2004), 'Einstein, Picasso', *Physics Education*, 39: 6, p. 484.

Nanotechnology Research Directions for Societal Needs in 2020, Springer: the Netherlands, vol. 1, pp. 1–28.

Niemetz, A. (2004), 'Singing cells, art, science and the noise in between', Master of Fine Art, University College Los Angeles.

Nordmann, A. (2004), 'Nanotechnology's worldview: new space for old cosmologies', *Technology and Society Magazine*, IEEE, 23, pp. 48–54.

Nykypanchuk, D., Maye, M. M. et al. (2008), 'DNA-guided crystallization of colloidal nanoparticles', *Nature*, 451: 7178, pp. 549–552.

Orban, C. (1997), *The Culture of Fragments: Word and Images in Futurism and Surrealism*, Amsterdam: Rodopi B. V.

Paul, C. (2007), 'The myth of immateriality: presenting and preserving new media', *MediaArtHistories*, O. Grau, MA: MIT Press.

Pelling, A. E., Sehati, S. et al. (2004), 'Local nanomechanical motion of the cell wall of Saccharomyces cerevisiae', *Science*, 305: 5687, 1147–1150.

Petrie, B. (1974), 'Boccioni and Bergson', *The Burlington Magazine*, 116: 852, pp. 140–147.

Phillips, M. (2010), 'A Mote it is ... art in the age of nano technology', C. University, Perth: The John Curtin Gallery.

Punt, M. (2002), 'Human consciousness and the postdigital analogue', *Leonardo*, 35: 2, pp. 119–120.

Raimondi, S. (2007), 'Nanoart: Out of this world, into the infinite', *Nanoart. Seeing the Invisible*, S. Raimondi (ed.), Milano: Skira.

Roberts, L. (2004), 'The death of the sensuous chemist: the "new" chemistry and the transformation of sensuous technology', *Empire of the Senses: The Sensual Culture Reader*, D. Howes (ed.), Oxford: Berg Press.

Roco, M. C., Hersam, M. C. et al. (2011), *The Long View of Nanotechnology Development: The National Nanotechnology Initiative at 10 Years*.

Serres, M. (1982), *Hermes: Literature, Science, Philosophy*, Baltimore: The John Hopkins University Press.

——— (2000), *The Birth of Physics*, Manchester: Clinamen Press Limited.

Shakespeare, W., *Hamlet*, Act 1, Scene 1, Line 129.

Simon, P. (1997), 'The virtualization of art practice: body knowledge and the engineering worldview', *Art Journal* 56(3): pp. 30–38.

Simondon, G. (1992), 'The genesis of the individual', in J. Crary and S. Kwinter (eds), *Incorporations*, New York: Zone Books, pp. 297–319.

Steenhuysen, J. (2008), 'DNA coaxed to build new gold crystals', *Reuters*.

Stephenson, N. (1995), *Diamond Age*, New York: Bantam Spectra Book.

Thacker, E. (2004, 2007), 'Networks, swarms, multitudes', from http://www.ctheory.net/articles.aspx?id=423.

Thomas, P. (2009a), 'Midas: a nanotechnological exploration of touch', *Leonardo*, 42: 3, p. 186.

——— (2009b), 'Reconfiguring Space', Saarbrücken, Germany, VDM Verlag.

Thomas, P. and Colless, E. (2010), 'The first international conference on transdisciplinary imaging at the intersections between art, science and culture, 2010', http://blogs.unsw.edu.au/tiic/welcome/.

Trust, T. V. S. (1979), Richard Feynman Video – Part 2: Fits of Reflection and Transmission – Quantum Behaviour. The Douglas Robb Memorial Lectures.

Vaughan, M. (2007), 'Introduction: Henri Bergson's "Creative Evolution"', *SubStance*, 36: 114, pp. 7–24.

Wang, J., Tetlow. G. A., Lucey, A. D., Armstrong, J. J., Leigh, M. S., Paduch, A., Sampson, D. D., Walsh, J. H., Eastwood, P. R., Hillman, D. R. and Harrison, S. (2006), 'Dynamics of the human upper airway: on the development of a three-dimensional computational model', *IFMBE Proceedings World Congress on Medical Physics and Biomechanical Engineering*, vol. 14, Springer-Verlag, Berlin.

## Acknowledgements

I would like to acknowledge the importance of all the people who have supported this publication. I would especially like to thank Kevin Raxworthy as a collaborator for all his thoughtful, technical and artistic input. In reference to the artists' projects presented I would like especially to thank Frederick De Wilde, Laurent Mignonneau, Ann Niemetz, Mike Phillips, Christa Sommerer and Victoria Vesna. I would also like to thank Oron Catts, Ionat Zurr, SymbioticA and Thomas Becker from the Nanochemistry Research Institute for their long standing support and assistance with my personal research projects.

I would like to express my gratitude to Edward Colless for his introduction as well as colleagues and friends Gail Kenning, Neil Sullivan, Jeremy Blank, Tim Gregory and David Eastwood who have assisted with editing and commenting on the work. I would also like to acknowledge the support of Bethan Ball from Intellect Books in assisting with this manuscript. Finally this book would not have been possible without the love, support and encouragement of my wife Christina Newberry.

# Index

## S



## T

## Author biography

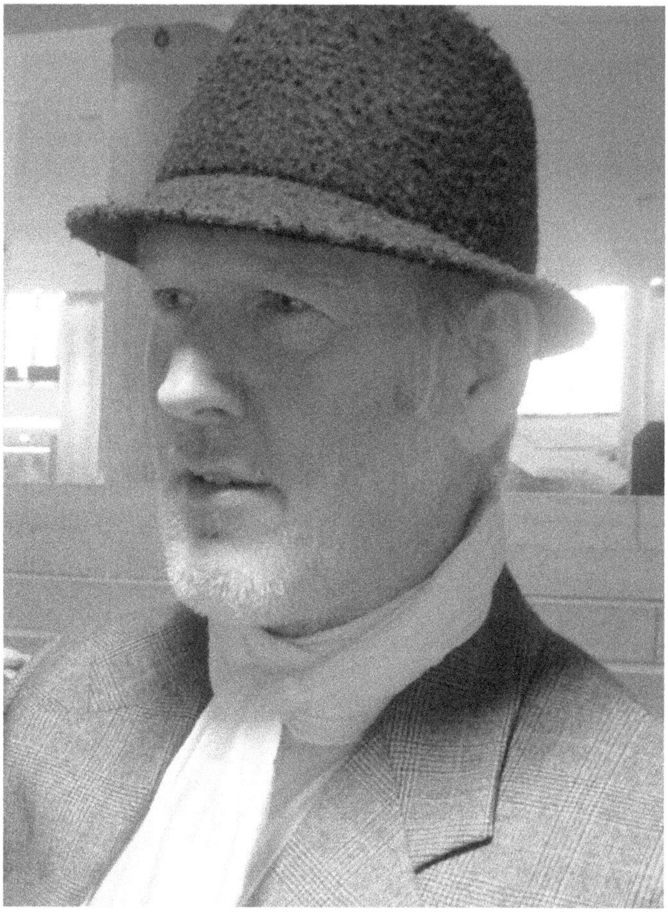

Dr Paul Thomas is an Associate Professor, Head of Painting at the College of Fine Art, University of New South Wales and was previously the instigator and coordinator of Collaborative Research in Art, Science and Humanity, (CRASH) Curtin University. Thomas has exhibited both nationally and internationally as well as chairing numerous international conferences. In 2000 Thomas was the founding Director and instigator of the Biennale of Electronic Arts Perth. Thomas's intention in writing this scholarly work about Nanoart is to explore ideas, thoughts and concerns from his experience as an artist, scholar and educator.

www.ingramcontent.com/pod-product-compliance
Ingram Content Group UK Ltd.
Pitfield, Milton Keynes, MK11 3LW, UK
UKHW051043120926
8556IPUK00031B/856

9 781841 507088